HANDBUCH DER ORIENTALISTIK

DRITTE ABTEILUNG

ERSTER BAND

LIEFERUNG 1

INDONESIAN CHRONOLOGY

HANDBUCH DER ORIENTALISTIK

Herausgegeben von B. Spuler
unter Mitarbeit von
H. Franke, J. Gonda, H. Hammitzsch, W. Helck, B. Hrouda,
H. Kähler, J. E. van Lohuizen-de Leeuw und F. Vos

DRITTE ABTEILUNG

INDONESIEN, MALAYSIA UND DIE PHILIPPINEN

UNTER EINSCHLUSS DER KAP-MALAIEN IN SÜDAFRIKA

HERAUSGEGEBEN VON H. KÄHLER

ERSTER BAND

GESCHICHTE

LIEFERUNG 1

INDONESIAN CHRONOLOGY

LEIDEN/KÖLN
E. J. BRILL
1978

INDONESIAN CHRONOLOGY

BY

J. G. DE CASPARIS

With 4 Plates

LEIDEN/KÖLN
E. J. BRILL
1978

ISBN 90 04 05752 8

PRINTED IN THE NETHERLANDS

CONTENTS

PREFACE

The main purpose of this book is to present a concise description of the chronological systems that have been in use in Indonesia through the ages. It was originally intended that this work should be partly devoted to the study of the concept of time in Indonesian civilizations, including an analysis of the significance of time in Indonesian historical writing and other literature. It soon appeared, however, that such a study, desirable though it may be, would considerably delay the publication of this section of the *Handbuch* on account of the serious problems which such a study would necessarily entail. On the other hand, the present author feels that there is no real need for a practical survey confined to a number of tables by means of which the reader can convert dates in Indonesian texts and documents into their Western equivalents. Such tables are readily available. Only a few have been added to the present publication for the sake of convenience. I therefore decided to present a descriptive analysis of Indonesian chronology in an historical context, incorporating both attempts at assessing the significance of time-reckoning in different periods of Indonesian history and some limited materials for practical use.

Apart from the philosophical issues, which deserve a special study, there remain at present no serious problems in Indonesian chronology. This favourable situation, standing in sharp contrast to other fields of Indonesian history and civilization, can largely be attributed to the fact that it has long been realized that a sound knowledge of chronology is a necessary prerequisite of the utilization of historical source materials. It is therefore not surprising that already the earliest Western historians of Indonesia, Raffles and his contemporaries, showed great interest in chronology. Great progress was achieved in the latter half of the nineteenth century, mainly owing to the enthusiasm of scholars like J. L. A. Brandes and G. P. Rouffaer. The former succeeded in calculating for the first time the precise equivalent of an Old Javanese date, while the latter composed the first comprehensive survey of Indonesian chronology in a long and detailed article in an encyclopaedia. In 1927 this article was brought up to date by three other scholars. Both Rouffaer's article and its later revision are still indispensable, the latter especially because it incorporates many interesting data concerning systems of chronology used in various parts of the Indonesian archipelago. Both articles also contain a number of useful tables which are, however, of little help for the study of the pre-Muslim chronology.

This desideratum has to some extent been compensated by further research on Indian chronology. Between 1923 and 1927 W. E. van Wijk published five articles which enabled also the non-mathematician and non-Indologist to utilize the chronological data of one of the most authoritative astronomical texts of ancient India, the *Sūryasiddhānta*, for the calculation of ancient dates. Although the tables in Van Wijk's articles are meant to apply only to the Indian subcontinent they can, with only minor adjustments, be used for the calculation of Indonesian dates. As most dates in inscriptions contain more details than those strictly necessary for the calculation of their Western equivalents, it is almost always possible to check the correctness of the results achieved by means of those tables.

The most spectacular progress in the field of ancient Indonesian chronology was, however, during the last twenty-five years achieved by the indefatigable efforts of L.-C. Damais, who by his own, admittedly rather laborious, methods succeeded in attaining the almost ultimate precision in finding the equivalents of the dates mentioned in old and later, Javanese as well as Balinese, inscriptions and manuscripts. These meticulous studies have brought additional benefits. Thus, they may in some cases confirm or allay the editor's suspicions about the inauthenticity of some documents. On the assumption that the authorities issuing official or semi-official documents would have ensured the correctness of all the details concerning their dates it follows that any incorrect date, for instance one in which a week-day is mentioned that does not tally with the day of the month, raises strong suspicion that the document is a falsification or else an incorrect copy of an authentic document.

While going through the proofs of the present publication I received a copy of M. C. Ricklefs' important and penetrating study of *Modern Javanese Historical Tradition*, which devotes large sections to modern Javanese chronology. In particular Appendix I: The Javanese Dating System (pp. 223-238) and Appendix II: Chronogram Words (pp. 239-244) contain a wealth of information which may supplement or replace the rather cursory treatment that this period receives in the present publication.

After all these important studies the present author feels that the most useful task at the present stage of our knowledge is that of collecting, shifting and summarizing the main results and presenting these in an historical setting.

Kew Gardens, January 1978

NOTE ON TRANSCRIPTION

In the transcription of Sanskrit and Old Javanese words and passages the methods used in other volumes of the *Handbuch* have been followed, in particular those of J. Gonda, *Old Indian* (1971) and F. H. van Naerssen and R. C. de Iongh, *The Economic and Administrative History of Early Indonesia* (1977). As, however, the methods adopted for Sanskrit and Old Javanese differ in a few details there may sometimes arise problems of how to transcribe Sanskrit words used in Old Javanese. Thus, Sanskrit words with *va* or *ṛ* are spelt with *wa* and *rĕ* (*ĕr*) if they are adopted in Old Javanese or occur in an Old Javanese context, whereas the Sanskrit transcription is followed in quotations from Sanskrit or in an Indian context. This dual method of transcription may lead to apparent inconsistencies in the spelling of words such as *Vaiśākha* or *Waiśākha* (name of the second month of the ancient Indian and Old Javanese year) and *vāra* or *wāra* (day of the week). The present author has preferred such inconsistencies to the use of the same system of transcription for both Sanskrit and Old Javanese, as it may sometimes render words difficult to recognize.

Finally, Old Javanese place names are spelt in the form in which they are actually found in inscriptions, transcribed according to the same methods as followed in the transcription of Old Javanese words, whereas modern names are spelt as they occur on maps, but with the modifications used in modern Indonesian spelling. Thus, Canggal and Dinoyo correspond to the older spelling Tjanggal and Dinojo or Dinaja.

ABBREVIATIONS

Acta Orient.	Acta Orientalia.
B.E.F.E.O.	Bulletin de l'École Française d'Extrême-Orient.
Bibl. Indon.	Bibliotheca Indonesica.
Bijdr. Kon. Inst.	Bijdragen tot de Taal-, Land- en Volkenkunde.
Contr. Ind. Soc.	Contributions to Indian Sociology.
Corpus Inscr. Indic.	Corpus Inscriptionum Indicarum.
Cult. Indië	Cultureel Indië.
Dinas Purb. Rep. Indonesia	Dinas Purbakala Republik Indonesia.
ed.	edition or edited by.
E.F.E.O.	École Française d'Extrême-Orient.
Encycl. Ned.-Indië	Encyclopaedie van Nederlandsch-Indië.
Ep. Ind.	Epigraphia Indica.
Epigr. Ind.	Epigraphia Indica.
Fasc.	Fascicule.
Fed. Mus. Journ.	Federation Museums Journal.
Feestb. Bat. Gen.	Feestbundel Koninklijk Bataviaasch Genootschap.
G.E.	Gupta Era.
Inscr. Ned.-Indië	Inscripties van Nederlandsch-Indië.
Journ. Am. Or. Soc.	Journal of the American Oriental Society.
Journ. Asiat.	Journal Asiatique.
Journ. Bombay Br. R.A.S.	Journal of the Bombay Branch of the Royal Asiatic Society.
Journ. Malay Br. R.A.S.	Journal of the Malay Branch of the Royal Asiatic Society.
Journ. Siam Soc.	Journal of the Siam Society.
Madj.	Madjallah (Majallah).
N.S.	New Series or Nouvelle Série.
O.J.O.	Oud-Javaansche Oorkonden.
Oud-Jav. Oork.	Oud-Javaansche Oorkonden.
Pras. Indon.	Prasasti Indonesia.
Publ. E.F.E.O.	Publications de l'École Française d'Extrême-Orient.
Rapp. Oudh. Dienst	Rapporten van den Oudheidkundigen Dienst.
R.A.S.	Royal Asiatic Society.
Tijdschr. Bat. Gen.	Tijdschrift van het Koninklijk Bataviaasch Genootschap.
Verh. Bat. Gen.	Verhandelingen van het Koninklijk Bataviaasch Genootschap.
Verh. Eerste Congres voor de T.L.V. van Java	Verhandelingen van het Eerste Congres voor de Taal-, Land- en Volkenkunde van Java.

CHAPTER ONE

INDONESIAN CHRONOLOGY BEFORE THE SIXTEENTH CENTURY

Some ideas of chronology, based on the ever-recurring cycles of days and nights, the phases of the moon and the seasons, are probably common to all mankind. In many societies, however, we find more or less sophisticated systems to fix past events in time and to determine the best time for events scheduled for the future. In some parts of Indonesia we find, in addition, from early times a keen awareness of the importance of chronology, at least in the major kingdoms. This awareness implies that it was felt that a literary text or a document had to belong to a definite point in the sequence of time and stood in a clear relationship to other, i.e. earlier, texts or documents. Time consciousness must have come naturally to societies in which maritime navigation had played an important part since remote times, as it involved some knowledge of astronomy and other means of orientation clearly associated with the measurement of time. On the other hand, agriculture and astrology, by defining or recommending the best times for certain activities, further stimulated interest in the reckoning of time.[1]

Such interest in chronology did not necessarily lead to the development of a strict system of chronology, but it contributed to the definition of various units of time in their interrelationships. The use of real systems of chronology is clearly associated with the emergence of major kingdoms, to which it has remained confined till relatively recent times. From the time of the earliest kingdoms, i.e. from the fifth or sixth century A.D., we have at our disposal a gradually increasing number of written sources which often supply information about chronology. Most of the extant inscriptions are dated, while the methods of dating tend to become more and more detailed and elaborate in the course of the centuries, at least before the fifteenth century. The value of these sources is, however, seriously limited by the fact that they are almost exclusively available for central and eastern Java from the eighth, as well as for Bali from the tenth century. There are, in addition, a few dated inscriptions from western Java, Sumatra and the Malay Peninsula, but none whatsoever from any of the other islands of the Indonesian archipelago.

[1] One of the most interesting of the modern works on the significance of chronology in an anthropological context is David F. Pocock, 'The Anthropology of Time-reckoning', *Contributions to Indian Sociology*, VII, 1964, pp. 18-29.

From the sixteenth century, on the other hand, the sources not only become much more numerous but are also spread more widely over the Indonesian archipelago. From that time, however, the systems that gain increasing influence are the Muslim and European chronologies, which require no discussion here, except for a few features that are characteristic of Indonesia.

The system of chronology which was in use before the fifteenth century is of Indian origin. As Indian chronology is not, or has not yet been, dealt with in this series, it is necessary to give a brief survey here as far as is useful as a background for the study of early Indonesian chronology. First of all, however, a brief survey of the little that is known of Indonesian chronology before the first contacts with India may not be out of place.

Earliest Indonesian chronology

As there are no known inscriptions or other texts predating the earliest contacts with India any description of Indonesian chronology during such periods must necessarily remain speculative. Nevertheless it is not impossible to draw a few likely conclusions with regard to the earliest Indonesian chronology.

Nearly a century ago one of the great founders of the study of Indonesia's past, J. L. A. Brandes, presented the first accurate calculation of an Old Javanese date and, in this connexion, called attention to some elements of dating which are unknown in India or anywhere else outside Indonesia and are therefore likely to be part of an old Indonesian heritage.[2] Among the most characteristic features of Javanese dating, as reflected in the inscriptions from the ninth century A.D., is the use of three different weeks of six, five and seven days. The inscriptions invariably record the names of the week-days of each of the three weeks on the date of the royal edict, and always in the order mentioned above, i.e. with the name of the day of the six-day week first. The names of these week-days, eighteen in all, must have been well known in Java, for they are not normally written in full but by abbreviations, actually the first syllables only. As to the names and their abbreviations the following list may be useful:[3]

[2] J. L. A. Brandes, 'Een rechterlijke Uitspraak uit het Jaar 927', *Tijdschr. Bat. Gen.*, XXXII, 1888, pp. 98-149.

[3] For further details see G. P. Rouffaer, 'Tijdrekening', *Enc. Ned.-Indië*, IV, 1905, pp. 445-460; D. G. Stibbe and C. Spat, m.m.v. E. M. Uhlenbeck, 'Tijdrekening', *ibid.*, V, Supplement, 1927, pp. 401-415; L.-C. Damais, 'Études d'Épigraphie Indonésienne', I. Méthode de Réduction des Dates Javanaises en Dates Européennes', *B.E.F.E.O.*, XLV, Fasc. 1, pp. 1-63, in particular, pp. 14-16.

Five-day Week	Six-day Week	Seven-day Week
Pahing (*pa*)	Tunglai (*tung*)	Āditya-wāra (*ā*)
Pon (*po*)	Haryang (*ha*)	Soma-wāra (*so*)
Wagai (*wa*)	Wurukung (*wu*)	Aṅgāra-wāra (*ang*)
Kaliwuan (*ka*)	Paniruan (*pa*)	Budha-wāra (*bu*)
Umanis (*u*)	Was (*wa*)	Brĕhaspati-wāra (*brĕ*)
	Mawulu (*ma*)	Śukra-wāra (*śu*)
		Śanaiścara-wāra (*śa*)

For some days of the seven-day week we occasionally find synonyms used, such as Candra-wāra (*ca*) for Soma-wāra and Suraguru-wāra (not found abbreviated) for Brĕhaspati-wāra and Śani-wāra for Śanaiścara-wāra. No synonyms are used of the days of the *ṣaḍ-* and *pañca-wāras*, but there are minor variations in spelling. In particular, the archaic *ai* of Tunglai and Wagai is often replaced by *e* and instead of Umanis we occasionally find Manis (*ma*) in some later inscriptions.[4]

It will be seen that some abbreviations are ambiguous, *viz. pa, wa* and, occasionally *ma*. As, however, the ambiguities do not apply to days in the same types of weeks and, in addition, the week-days are invariably mentioned in the same order with the day of the six-day weeks first, no misunderstanding can arise.

In addition to the three different types of weeks mentioned above there also exists a three-day week, which is attested in Bali from the tenth century. The names are: Pasah (*pa*), Beteng (*be*) and Kajeng (*ka*).[5]

It is of interest to note that all the names, except those of the *sapta-wāra*, are Indonesian. They are also etymologically obscure except for a few that can be 'translated' (thus, *Paniruan* can be explained as 'example to be imitated', from the root *tiru*; *was* could mean 'clear' or 'bright').[6] Nothing can therefore be concluded about the origin of these

[4] For variants and other details concerning the weekdays see Damais in the article mentioned in the previous note.

[5] These are all modern names; see W. O. J. Nieuwenkamp, 'Een Balineesche Kalender', *Bijdr. Kon. Inst.*, LXIX, 1914, pp. 112-126, who gives the names Doya, Waya, and Byantara (Dwāra, Wāhya, and Abhyantara), whereas the inscriptions in Old Balinese give Wijayapura, Wijayakrānta and Wijayamaṇḍala, preceded by the words *rggas* (i.e. phonetically *rĕgas*) *pasar*; cf. R. Goris, *Prasasti Bali*, II, 1954, p. 297, s.v. *rggas*. The oldest known example is in an inscription from Sukawana dated A.D. 883 (R. Goris, *op. cit.*, I, p. 54, line III a 1/2). For the significance of these names see W. F. Stutterheim, *Oudheden van Bali*, I, 1929, pp. 80-85.

[6] Some other possible etymologies have been suggested by Rouffaer in the above-mentioned encyclopaedia article, as well as by Stibbe and Spat in their revision of 1927. Some of the proposed etymologies are quite unlikely. Thus, *pon* is explained as *pu-an*, derived from *pu*, 'lord' etc. with the suffix *-an*, but in that case one would have expected to find *puan* and *pwan* in Old Javanese. But the Old Javanese inscriptions invariably give *pon*, abbreviated *po*, which suggests that the *o* has developed from an old diphthong *au*, just as Old Javanese *ron*, 'leaf', corresponds to *daun* in Malay and Indonesian. For *kaliwon* Rouffaer wrongly mentions an older form *kliyon*, but Old Javanese always gives *kaliwuan*.

names. Although the names are unattested outside Java and Bali, there is reason to believe that they are very old. Their use is always consistent, and no mistakes are made. They are also regarded as an essential element of the date, for they are never omitted. The days of the *saptawāra* are, on the contrary, all Sanskrit, so that one might be inclined to trace the origin of the seven-day week back to India, where the names are, in fact, attested earlier, though not very much earlier,[7] than in Java. Yet, it would be surprising if the seven-day week was unknown in Indonesia before the first contacts with India for, unlike weeks of five and six days, it reflects a natural division of time as its length corresponds to the interval between each of the phases of the moon. The fact that the names of the week-days are always given in Sanskrit is not necessarily a strong argument in favour of their Indian origin. Thus, in modern Javanese (and Indonesian) the names of these same weekdays are derived from Arabic ordinal numbers, adapted to Indonesian[8], while the old Sanskrit names have almost been forgotten. The Arabic names were clearly substituted for the old Sanskrit ones at the time when Islam became the prevailing religion. The old Sanskrit names have not completely disappeared but are occasionally found in literary Javanese, but this is no doubt a consequence of the strength of the learned literary tradition in Java. As there is no evidence for any literary tradition predating the earliest references to the seven-day week (the earliest attested example is in the Canggal inscription of A.D. 732) any earlier Javanese names for the weekdays would have been lost even if they ever existed.

One particular argument in favour of the antiquity of the seven-day week can be based on the names of the *wukus*, i.e. the names given to each of the seven-day weeks. There are thirty *wukus* in all, which yields a cycle of 210 days. The thirty *wuku* names, attested from the tenth century A.D., are nearly all clearly Indonesian in origin. There are two doubtful names. The first *wuku*, Sinta, may be named after Sītā, but this is by no means certain, while No. 25, Bala, may represent the well

[7] The earliest known evidence for the use of the seven-day week in ancient India is the Eran Stone Pillar Inscription of Budha Gupta, dated G.E. 165, corresponding to A.D. 484; the first half of verse 2 reads: *āṣāḍhamāsaśukladvādaśyāṃ suraguror=divase*, 'On the twelfth lunar day (*tithi*) of Āṣāḍha in the bright half of the month, on the day of the Teacher of the Gods', i.e. Bṛhaspativāra, 'Thursday' (J. F. Fleet, *Corpus Inscr. Indic.*, III, 1887, reprint Varanasi, 1970, pp. 88-90; D. C. Sircar, *Select Inscriptions bearing on Indian History and Civilization*, I, 2nd ed., 1965, pp. 334-336). There is, however, literary evidence to suggest that the use of the seven-day week in ancient India may be several centuries older, from around the beginning of the Christian era; cf. P. V. Kane, *History of Dharmaśāstra*, V, Part I, 1958, pp. 676-685.

[8] For Sunday, however, the term (*Hari*) *Minggu*, derived from Portuguese *Domenico*, is used by the side of *Akad* in Javanese and *Ahad* in Indonesian, both derived from Arabic أَحد, 'first'.

known Sanskrit word for 'force, army' etc. Neither etymology is beyond doubt. Even if both are accepted as words of Sanskrit origin their number is too small to be significant, as the two names could be later substitutes for older Indonesian names. On the other hand, it may be argued that the antiquity of the *wuku* system is itself doubtful since these names are attested not earlier than the tenth century.[9] Although it is theoretically possible that the *wuku* system should have been 'invented' at a comparatively late stage this does not seem likely at a time when dates had already become quite elaborate by the addition of a number of astronomical details of Indian origin. Most of the *wuku* names have no other meaning but that of indicating a particular *wuku* and many of these are etymologically obscure. This would by no means be surprising for names handed down from the remote past but would seem quite unlikely for names given at a late stage.

In conclusion, it seems likely that the entire system of three types of weeks and the 210 day cycle is of Indonesian and, in particular, Javanese origin. It seems to reflect the importance of cycles of 210 days, a period without clear astronomical significance. Its use does not necessarily exclude the use of 'natural' units of time such as lunar months and solar years. On the contrary, the fact that the words for 'month' and 'year' (*bulan* and *tahun*) are Austronesian shows that these concepts go back to a remote antiquity, but there is no indication that the different units of time had already been combined into a coherent system, as we find it used from the seventh century A.D. or earlier. This brings us, however, to a period when contact with the Indian subcontinent had since long been established. The system used from that time in Sumatra and Java, as well as in Bali from a few centuries later, is in fact of Indian origin but with the inclusion of some Indonesian elements, in particular the use of the above mentioned weeks. Before proceeding to a description

[9] The oldest mention of a *wuku* is in an inscription of king Wawa of A.D. 928; see Damais, *B.E.F.E.O.*, XLV, Fasc. 1, 1951, p. 7. Most important details concerning the *wukus*, with lists of Old Javanese, modern Javanese and Balinese names and their variants are fully discussed by Damais in the same article. In this connexion Damais calls attention to the fact that all the early examples of the use of *wuku* names are from eastern Java, while the earlier central Javanese inscriptions do not give any *wuku* names. He then concludes to an East Javanese origin of these names. It should however, be added that *wuku* names are also absent from the few known East Javanese inscriptions dated before 928, such as the Ponorogo copper-plate inscription of A.D. 901 (Brandes, *O.J.O.*, No. XXIII; Sarkar, *Corpus*, II, No. LXI; Damais, *B.E.F.E.O.*, XLVI, 1, 1952, p. 44, No. 68; *B.E.F.E.O.*, XLVII, 1, 1955, pp. 40 f.), and the Singosari stone inscription of 915 (Brandes, *O.J.O.*, No. XXX; Damais, *B.E.F.E.O.*, XLVI, 1, 1952, p. 52, No. 95; *B.E.F.E.O.*, XLVII, 1, 1955 pp. 50 f.). As both inscriptions have elaborate dates, including the names of *nakṣatra*, *dewatā* and *yoga*, the absence of *wuku* names seems significant. It therefore appears likely that the mention or omission of *wuku* names does not depend on the area, but on the period: only from 928 did the *wuku* names become part of the official protocol.

of the system of chronology as used in parts of Indonesia from the seventh century A.D. it will therefore be necessary to present a brief outline of Indian chronology, especially as no volume in this series has hitherto appeared on this field.

Indian Chronology

The survey which follows is mainly concerned with those aspects of Indian chronology that are important for the understanding of Indonesian developments.

Although it has often been alleged that the ancient Indians were too strongly preoccupied with speculation about the absolute and eternity to be interested in something as temporary as time itself, it soon appears that the opposite is true. Indian religion and philosophy abound in theories about time, also conceived of as the Divine Destroyer (*Kāla*). Speculation about time leads, especially in the Epics and the Purāṇas,[10] but already foreshadowed in Vedic literature, to the conception of immensely long periods such as *kalpas* and *yugas* as well as of fragments of seconds. It is true that the figure given for periods, and also for the lengths of reigns of legendary kings, appear imaginary and often phantastic, but the fact that such figures are invariably given attests to the importance attached to time.[11]

Apart from the religious and philosophical importance of time, the Indians, living in well-ordered societies since time immemorial, required more or less precise definitions of points of time and of time intervals. As all measurement of time is necessarily linked with the movements of heavenly bodies there has always been a close relationship between astronomy and chronology.

As Indian influence in South East Asia made itself felt at a stage when Indian culture was, after long developments for twelve centuries or longer, reaching its classical stage, there is no need to discuss the early history of Indian chronology and astronomy.

Indian chronology in the classical age, i.e. approximately from the

[10] One of the most useful publications on ancient Indian chronology is W. E. van Wijk, *Decimal Tables for the Reduction of Hindu Dates from the Data of the Sūrya-Siddhānta*, The Hague, 1938. This work is based on the very elaborate articles on Hindu chronology published in *Acta Orientalia* between 1921 and 1926. Useful introductory surveys are given in L. Renou and J. Filliozat, *L'Inde Classique*, I-II, 1947-53, and in A. L. Basham, *The Wonder that was India*, 1954 (reprints 1956 and 1961), especially Appendix III, 'The Calendar', pp. 492 ff. One recent work by R. Billard, *L'Astronomie Indienne*, *Publ. E.F.E.O.*, LXXXIII, 1971, contains much important material for the bases of Indian chronology.

[11] The most penetrating modern study on time and time-reckoning in ancient India, including its philosophical foundations and its religious implications, is in P. V. Kane's *History of Dharmaśāstra*, Vol. V, Part I, pp. 463-718.

fourth to the seventh century A.D. has been studied with great detail. Among the most valuable contributions to this topic are those by the Dutch scholar W. E. van Wijk.[12] Some of the results of these studies are contained in a number of tables based on the data of the *Sūryasiddhānta*, one of the most authoritative texts dating back to probably the fifth or sixth century A.D.[13] Other authoritative texts, such as the *Āryasiddhānta* and the *Brahmasiddhānta*, though often more accurate in minute details, mainly agree with the data of the *Sūryasiddhānta*.[14]

As in most other chronological systems the basic problem of Indian chronology was that of reconciling the lunar (i.e. synodic) month with the solar year, as the latter is approximately eleven days longer than twelve lunar months. Another, somewhat less serious, problem is that of reconciling both with the mean length of a day, counted from sunrise to sunrise, which is slightly longer than 1/30th of a lunar month and very slightly longer than 1/365th of a solar year. The principal mechanism by which adjustment between the synodic months and the solar years was achieved was the intercalation of an extra month approximately seven times in nineteen years. In principle, a month was 'repeated' (by the use of the same name, but with the addition of a word meaning 'second') whenever two synodic months fell within one so-called solar month, which is precisely 1/12th of a solar year or approximately 30.438 days, whereas the length of the synodic month is about 29.531 days. To determine the beginning of months two different systems were used: the *pūrṇimānta* system according to which the months ended (and, of course, started) which each full moon and the *amānta* system in which the months were counted from one new moon to the next. In general, the former system was followed in northern India, the latter in southern India, although the actual data are more complicated. According to the *amānta* system, which is that which was exclusively used all over South East Asia, there was to be a repetition of any month that started with a new moon due to take place less than 30.438 *minus* 29.531, i.e. .907

[12] W. E. van Wijk, 'On Hindu Chronology', *Acta Orient.*, I, 1921, pp. 206-233; II, 1922, pp. 55-62; II, 1922, pp. 235-249; IV, 1925, pp. 55-80; V, 1927, pp. 1-28.

[13] It is thought that the 'original' *Sūryasiddhānta* may date back to the fourth century A.D. In its present form, however, it may be somewhat later, though its author must be earlier than Varāhamihira, who is usually assigned to the beginning of the sixth century A.D. Cf. the detailed discussion by J. Filliozat in L. Renou and J. Filliozat, *L'Inde Classique*, II, 1953, pp. 180-187, and Billard, *op. cit.*, pp. 153-155.

[14] W. E. Clark, *The Āryabhaṭīya of Āryabhaṭa*, Chicago, 1930. Brahmagupta's *Brahmasphuṭasiddhānta* with its commentary *Khaṇḍakhādyaka* has been edited by P. Sengupta, Calcutta, 1941, and translated by Jyotishacharyya, Calcutta, 1925. For the problems of dating the reader is referred to the publications by Filliozat and Billard mentioned in the previous note. Brahmagupta's date (beginning of the seventh century A.D.) is well established.

days or twenty-one hours, 44 minutes and 38,04 seconds after mean sunrise of the first day of the month.[15]

It is likely that the precise time of each new moon was not determined by observation, as was the case in Muslim chronology, but by previous calculation on the basis of tables. One of the numerous great merits of Van Wijk in this field has been that of providing us with precise tables of the Western equivalents of the figures presented in the *Sūryasiddhānta*, thus enabling us to find the equivalent Western (i.e. Julian) dates of those expressed in Indian inscriptions and texts—at least, whenever, the details are preserved in their complete and correct form. The beginning of the year, i.e. the first day of the bright half of the month (the period from the new to the full moon) of Caitra, was fixed as the date of the first new moon following *Mīnasaṃkrānti*, the entrance of the sun into the Fishes, approximately one month before the vernal equinox. Whenever the first new moon was due within 10,872 days after *Mīnasaṃkrānti* the year would be one of thirteen months; the actual month due to be repeated depended on the precise interval between *Mīnasaṃkrānti* and the next new moon. Thus, Caitra, the month following *Mīnasaṃkrānti*, was to be repeated if the interval was smaller than .907; Vaiśākha, the second month, if between .907 and 1.812, Jyaiṣṭha, the third, if between 1.812 and 2.178 etc.

From this survey it may have become clear that Indian chronology not only counts with solar years (as reflected in, for instance, the date of the *Mīnasaṃkrānti*) but also with 'lunar years' of twelve or thirteen months. Similarly, by the side of the normal synodic lunar month there is also a 'solar month' equal to 1/12th of a solar year. The same contrast is observed for the concept of the day, because in addition to the normal solar day, reckoned from (mean) sunrise to the next sunrise, there is also a 'lunar day' (*tithi*) equal to 1/30th of a (synodic) month, i.e. about .9844 days or 22.464 minutes shorter than a normal day. It therefore happens about once in every two months that two *tithi*s begin within the same solar day, with the result that the second *tithi* is 'lost' or expunged.

---

[15] These figures are all based on Van Wijk's tables (see note 10 above). There are unfortunately too few inscriptions dated in intercalary months to enable us to ascertain the precise system followed in ancient Indonesia. Calculations based on texts other than the *Sūryasiddhānta* may yield slightly different figures, but, except for a few rare marginal cases, the final result would be the same. We also do not know whether the calculations were based on true or mean sunrise, but the latter seems more likely. If it was determined by the means of tables (the inscriptions had to be prepared well in advance of the date given at their beginning, for the latter marks the time when the formal consecration of the land and its transfer to the donee(s) were to take place), it is probable that the Indian data were adapted to use in areas situated about 40 degrees east of 'Ujjayini', the theoretical basis of ancient Indian chronology.

The *tithi* may appear a somewhat artificial time unit, yet it came into general use in dating in India and it is invariably mentioned in the Indonesian inscriptions immediately after the name of the months. The thirty *tithi*s are, however, divided into two groups of fifteen, the first during the waxing moon (*śukla-pakṣa*), the second during the waning moon (*kṛṣṇa-pakṣa*).[16]

In this connexion attention may be drawn to another apparently artificial unit of Indian chronology which found its way to ancient Indonesia: the *nakṣatra*, 'lunar mansion', each of the twenty-seven or twenty-eight constellations into which the apparent path of the moon (ecliptica) is divided. In Indonesia only the system of twenty-seven *nakṣatra*s seems to have been used. The length of one *nakṣatra*, as a time unit, is therefore equal to 1/27th of the time required by the moon to return to the same point at the ecliptica, i.e. a sidereal month of 27.322 days.

The Śaka Era

Many different eras were used in ancient India. Some of these were religious in origin, such as the era reckoned from Lord Buddha's Nirvāṇa in 544 B.C. according to the tradition followed in Sri Lanka from the thirteenth century or earlier; others, such as the Gupta era of A.D. 320, were dynastic, dating back to the beginning of the reign of Candra Gupta I or to an important event at or near the beginning of his reign. It is, however, striking that the most widely used eras of ancient India, viz. the Vikrama era of 57 B.C., the Śaka era of A.D. 78 and the Kalacuri era of A.D. 248, are of uncertain origin. Both the Vikrama and the Śaka eras are, according to legend, based on the struggles for the city of Ujjayinī (Ujjain) in present Malwa, Madhya Pradesh. The former would record the conquest by the Śātavāhana king Vikramāditya of Ujjayinī, which had been taken by the Śakas (Scythians), the latter its reconquest by the Śakas. Most modern scholars, though often retaining some features of the legend, assign the Vikrama era to one of the Indo-Scythian or Indo-Parthian kings, such as Azes I or Vonones, and the Śaka era to one of the great Kuṣāṇas, either Vīma Kadphises or, more often, Kaniṣka I. In most cases it is tacitly or explicitly assumed that both eras started

[16] Some *tithi*s have special names, such as *pratipāda* for the first of the *śuklapakṣa*. For the others Sanskrit ordinal numbers were used (always in the feminine, as *tithi* must be understood). For the details of spelling etc. in Old Javanese and Balinese cf. Damais in *B.E.F.E.O.*, XLV, Fasc. 1, 1951, p. 13. Damais thought that *tithi* lost its original meaning of 'lunar day' (i.e. precisely one thirtieth of a lunar month) and came to be used to indicate a (civil) day. Although this certainly happened at some stage, there is no evidence to show that this was the case in ancient Indonesia. It seems more likely that *tithi* lost its precise meaning at a time (after the ancient period) when the intricacies of the Indian system of time-reckoning were no longer known and applied.

from counting by regnal years of some of the above mentioned (sometimes also other) kings but developed into regular eras when their successors, instead of counting back to their own consecration, continued the reckoning of their predecessors.[17]

Of all the above mentioned Indian eras only the Śaka era spread to Indonesia. Also in Mainland South East Asia it was by far the most commonly used era in the early period. Thus, it was invariably used in the numerous inscriptions of Cambodia and Campa. The reasons for its popularity are, however, difficult to explain. It has often been pointed out that the Śaka era, which was rarely used in northern India until comparatively late, was from the sixth century A.D. widely used in the inscriptions of the western Deccan. In general, the Deccan is one of the Indian regions which generated most of the Indian influence in South East Asia. It should, however, be added that it seems as though such influence came from the eastern part of the peninsula, from Orissa and Andhra Pradesh down to Tamilnāḍu and Sri Lanka, whereas elements pointing specifically to the western Deccan are difficult to find. One can even argue that, on the contrary, some of the characteristic religious currents of the western Deccan, notably Jainism and Vīraśaivism, do not appear to have exercised any influence in South East Asia. In addition, the linguistic influences from southern India all point to the Tamilnāḍu,[18] and no words have yet been traced back to languages as Kannaḍa (Canarese) or Malayalam. It would be difficult to explain why, in that case, only the Śaka era, but none of the other aspects of the culture of the western Deccan, should have spread to South East Asia.

For this reason it seems that a different explanation is required. In addition to its regional connexions with parts of western India the Śaka era was also *the* era used in all astronomical works, usually in conjunction with the Kaliyuga era of 3102 B.C. It has therefore been called '*l'ère des astronomes*' in one of the most recent works on Indian astronomy.[19] It is well known that, in general, Indian influence in South East

---

[17] There is a vast literature on the origin of Indian eras. For the up-to-date accounts and discussions see especially J. E. van Lohuizen-de Leeuw, *The Scythian Period*, 1949, pp. 1-72 and A. L. Basham (ed.), *Papers on the Date of Kaniṣka*, Leiden, 1968.

[18] The Tamil loan words in Malay (and Indonesian) were discussed by Ph. S. van Ronkel in several articles, especially 'Het Tamil-element in het Maleisch', *Tijdschr. Bat. Gen.*, XLV, 1902, pp. 97 ff. Cf. also J. Gonda, *Sanskrit in Indonesia*, 2nd ed., Nagpur, 1973, pp. 161-166. As far as I know, no words in Indonesian languages have been traced back to Dravidian languages other than Tamil. Although they may well exist, they are probably so few in number that they will not affect the argument.

[19] R. Billard, *op. cit.*, p. 123.

Asia was almost exclusively of the learned type, the culture of the brahmins, especially the paṇḍits. It is significant that the linguistic influences in South East Asia are almost exclusively Sanskritic and, in mainland South East Asia in a comparatively late period, Pali, which is no less than Sanskrit a learned and literary language. It is true that a small number of Tamil inscriptions have been found in South East Asia. It should, however, be noted that wherever the texts of the inscriptions have remained legible they were the work of Tamil settlers in different parts of South East Asia, notably at Lobok Tua (Barus, northern Sumatra, Indonesia), Pra Narai (Takuapa, southern Thailand), and Myinpagan (near Pagan, Burma).[20] However interesting the presence of such Tamil settlements may be, it would appear that their influence upon the South East Asian communities among which they lived was strictly limited. In these circumstances it is not altogether surprising that we find no clear evidence for any influence from the popular eras used by these settlers but, on the contrary, from the 'scriptural' eras, such as those used by the Indian astronomers, above all the Śaka era of A.D. 78. Other learned eras, such as the above-mentioned Kaliyuga era and the so-called Laukika or Saptarṣi Era,[21] were not practical. The former was not completely unknown in South East Asia but, as in India, never really became accepted, partly probably owing to the need for very high figures in the three and four thousands. The Laukika era was even less useful as it normally omitted the hundreds in the dates. Its utility was therefore confined to relatively short periods.

The above discussion may not provide us with a complete and satisfactory explanation of the popularity of the Śaka era in South East Asia. It may, however, emphasize the most prominent feature of Indian influence in South East Asia: its close association with the learned

[20] K. A. Nilakanta Sastri, 'A Tamil merchant guild in Sumatra', *Tijdschr. Bat. Gen.*, LXXII, 1932, pp. 314-329; 'Takuapa and its Tamil Inscription', *Journ. Mal Br. R.A.S.*, XXII, 1, 1949, pp. 25-30; A. Lamb, 'Miscellaneous Papers on Early Hindu and Buddhist Settlements in Northern Malaya', *Fed. Mus. Journ.*, N.S., VI, 1963, pp. 1-90; E. Hultzsch, 'The Tamil/Sanskrit Vaiṣṇava Inscription, 13th century, found near Myinkaba village, Pagan', *Epigr. Ind.*, VII, 1902, pp. 197 f.

[21] The starting point of the Laukika or Saptarṣi era is the first of the bright half of Caitra of Kali Yuga expired 25 = 3076 B.C. It was used quite regularly in ancient Kashmir and adjoining areas of the Punjab also for most of the dates in the famous chronicle of Kashmir; the *Rājataraṅgiṇī*. See G. Bühler, 'Detailed Report of a tour in search of Sanskrit MSS in Kaśmir', *Journ. Bombay Br. R.A.S.*, Extra No., 1877, pp. 59 ff., quoted by M. A. Stein, *Kalhaṇa's Rājataraṅgiṇī*, 1900, reprint, Delhi, 1961, I, pp. 58 f. As this era starts precisely a quarter of a century after the commencement of the Kali Yuga era there is an obvious relationship between the two. As to the latter, it was widely used in astronomical texts but not so frequently, and only in the early medieval period, for inscriptions in India. In South East Asia there is an isolated example of its use in the Ch'aya inscription of Candrabhānu, dated K.Y. 4332 (expired: *kaliyugabarṣāṇi dvātriṅśādhikas=trīṇi śatādhikacatvārasahasrāṇi*); see G. Cœdès, 'Le Royaume de Çrīvijaya', *B.E.F.E.O.*, XVIII, 1918, Fasc. 6, p. 32.

Pan-Indian tradition in contrast to the numerous local traditions, whether popular or dynastic in nature.

Even within the learned Pan-Indian tradition, however, the Indonesians, like the élites of other parts of South East Asia, appear to have been selective. Practical considerations no doubt accounted for the absence of the 'theoretical' eras (Kaliyuga, Laukika etc.). The absence of the Vikrama and Kalacuri eras is more difficult to explain.[22]

Regnal Years

Dating in regnal years of the king issuing the inscriptions was common in India, not only in the oldest period (such as the inscriptions of Aśoka) but in some parts of the subcontinent (e.g. among the Pālas of Bengal and the Cōḷas of South India) till the twelfth century. In Indonesia I am aware of only one single example of dating in regnal years, viz. in the Tugu inscription of Pūrṇavarman issued in the 'twenty-second year of his prosperous reign'.[23] This is in one of the oldest inscriptions which, also in other respects, appears different from other Indonesian inscriptions in its orientation. Even in cases where inscriptions list achievements of a king in chronological order, as in the bilingual 'Calcutta' inscription of Airlangga (dated A.D. 1041), this is not done with reference to the king's regnal years but in absolute dates of the Śaka era.[24] I am unaware of any examples of a similar procedure in Indian or Sinhalese inscriptions, in which such enumerations of achievements are given in regnal years, e.g. in the Hāthigumphā inscription of Khāravela, or even without any chronological frame. Thus, the Allahabad Pillar inscription of Samudra Gupta [25] of the fourth century A.D. and the Aihōḷe inscription of Pulakeśin II give no indications whatever, and it is by no means certain, if not unlikely, that the achievements are listed in chronological order.[26] In this respect the emphasis on absolute chronology was at

[22] As to the Vikrama era of 57 B.C., known under different names such as Kṛta or Mālava era before the eighth or perhaps ninth century (if the Dhiniki, Kathiawar, inscription of V.S. 794 is spurious), see the excellent account by D. C. Sircar in *Indian Epigraphy*, Delhi, Varanasi and Patna, 1965, pp. 251-258. The Vikrama era was in wide-spread use in northern and western India, the Kalacuri era originally also in western India but later especially among the Kalacuris of the Jabalpur area (V. V. Mirashi, 'Inscriptions of the Kalachuri-Chedi Era', *Corpus Inscr. Indic.* IV, Parts I and II, 1955). Neither era was ever popular in South India, a fact which may to some extent account for the complete absence of both in South East Asia.

[23] J. Ph. Vogel, 'The Earliest Sanskrit Inscriptions of Java', *Publ. Oudh. Dienst*, I, 1925, p. 35.

[24] H. Kern, 'De steen van den berg Pĕnanggungan (Surabaya), thans in 't Indian Museum te Calcutta', *Verspr. Geschr.* VI, 1917 (original articles of 1885 & 1913), pp. 83-114. The Sanskrit verses give the years in chronograms, whereas the Old Javanese prose uses numerals.

[25] J. F. Fleet, 'Inscriptions of the Early Gupta Kings and their Successors', *Corpus Inscr. Ind.*, III, 1887, reprint Varanasi, 1970, No. I, pp. 1-17.

[26] F. Kielhorn, 'The Aihole Inscription of Pulakeśin II', *Epigr. Ind.*, VI, 1898, pp. 1-12.

least as strong in ancient Indonesia—i.e., in ancient Java, Sumatra and Bali—as in mainland South East Asia. Thus, in the great Sukhodaya (Sukhot'ai) inscription of Rāma Gaṃheng no regnal years are used in referring to the king's achievements but absolute dates in the Śaka[27] era. Apart from 1214 (i.e. A.D. 1292), the date of the inscription, we also find 1207 (A.D. 1285) and 1205 (A.D. 1283). It is curious that, in the last mentioned case, the three dates are given in this order, not in their proper chronological sequence.[28] In other cases, however, for instance in the great inscription of Sdok Kak Thom,[29] the strict chronological order is adhered to.[30]

It may be proper, in this connexion, to call attention to a few other details of Indian chronology which found no place in the Indonesian system. The so-called Jovian years with cycles of either twelve or sixty years,[31] based on the movements of the planet Jupiter (Bṛhaspati), has apparently never been used in Indonesia, although it was quite popular in the Deccan. On the other hand, the eight-year cycle of Javanese chronology, attested from the eighteenth century, was apparently unknown in India.[32] Reckoning in current (*vartamāna*) years, popular in parts of India during certain periods, the use of solar (*saura*) months or of *pūrṇimānta* months[33] are other elements of Indian chronology which were not adopted in Indonesia.[34] In this, as well as in many other respects the

[27] G. Cœdès, *Recueil des Inscriptions du Siam*, I, 1924, pp. 37-00; A. B. Griswold and Prasert ṇa Nagara, 'The Inscription of King Rāma Gaṃheṅ of Sukhodaya', *Journ. Siam Soc.*, LIX, 1971, pp. 179-228.

[28] The explanation is no doubt, as the editors of the last mentioned article of note 27 suggest (*ibid.*, pp. 194 f.), due to the fact that the Śaka dates 1207 and 1205 are both found in 'Epilogue' I, which is retrospective in character.

[29] L. Finot, L'inscription de Sdok Kak Thom', *B.E.F.E.O.*, XV, 1915, pp. 53-106; G. Cœdès and P. Dupont, Les stèles de Sdok Kak Thoṃ, de Phnoṃ Sandak et de Prah Vihar', *B.E.F.E.O.*, XLIII, 1943-46, pp. 134-154.

[30] Unlike the inscriptions of Indonesia those of Cambodia regularly mention the year of accession at least in the longer Sanskrit inscriptions. Thus, in the stele of Tuol Prasat (G. Cœdès, *Inscriptions du Cambodge*, II, 1942, pp. 97-114), dated A.D. 1003, king Jayavīravarman is said to have acceded to the throne the previous year: *avdhidvirandhrādhigatādhirājyah*, 'parvenu à la royauté suprême en (9) ouvertures—deux—(4) océans', i.e. (Śaka) 924 (in contrast to Indonesian chronograms, which have always to be read from the smallest to the largest units, the Cambodian chronograms follow the order in which the figures are written).

[31] J. F. Fleet, 'Inscriptions of the Early Gupta Kings and their Successors', *Corpus Inscr. Ind.*, III, 1887, reprint 1970, Introduction, pp. 101-124; D. C. Sircar, *Indian Epigraphy*, 1965, pp. 267-269.

[32] The eight-year *windu* cycle of pre-modern Java will be discussed towards the end of this article.

[33] The concept of solar months must, however, have been known in ancient Indonesia because the system of repeating the name of the previous month whenever there is a second new moon within one (mean) *saṃkrānti* (one-twelfth of a solar year) was regularly applied. *Pūrṇimānta* months (starting and ending at the time of full moon), though usual in many parts of ancient India, appear to be totally unknown in Indonesia.

[34] Many other details can be added to this list, such as years starting in Vaiśākha or in Kārttika, the counting of months by seasons etc.

Indonesians were selective, choosing whatever seemed useful or appropriate in view of Indonesian needs and customs. Wherever Indian tradition offered different alternatives a choice was made and subsequently firmly adhered to. It is, however, difficult to determine why elapsed years were preferred to current years, luni-solar (i.e. synodic) months to solar months, or *amānta* to *pūrṇimānta* months. In general it seems as though South Indian tradition carried relatively great influence. The chronological concepts are generally consistent with this view, but there arise difficulties as soon as one tries to define more precisely which area of South India played the major part. Thus, while most of the early Indian influences in Indonesia can be traced back to the Pallava kingdom it should be emphasized that the Śaka era was only rarely used by the Pallavas and in relatively late inscriptions. The Pallavas, like the Cōḷas and the Pāṇḍyas, nearly always used regnal years, the last in a still imperfectly understood manner.[35] On the other hand, there is early evidence for the use of the Śaka era in the western Deccan, in particular in the inscriptions of the Western Cālukyas of Vātāpi (Badami) and the Western Gāṅgas of Mysore. The earliest known inscription in the Śaka era, in which the era is in fact thus indicated, is the Badami inscription of the Cālukya king Vallabheśvara, dated Śaka 465 or A.D. 543,[36] although there is an earlier example found in a text, the Jain *Lokavibhāga*,, dated Śaka 380 or A.D. 458 in the 22nd regnal year of the Pallava king Siṃhavarman (I).[37] This example demonstrates that the Śaka era, though not used in inscriptions until centuries later, was known and used in South India about one and a half centuries earlier than its first known appearance in South East Asia. This same example lends strong support to the view expressed by D. C. Sircar that the early expansion of the Śaka era in the Indian subcontinent reflects the expansion of Jainism and, especially, the influence exerted by individual 'Jain scholars' who were 'apparently employed as astronomers and administrators at the courts of the various South Indian royal families.' [38]

[35] Many Pāṇḍya inscriptions give two dates, one 'opposite' the other. Most scholars add up the two figures, the sum of which they regard as the king's regnal years (D. C. Sircar, *op. cit.*, pp. 241 f.; K. A. Nilakanta Sastri, *A History of South India*, 3rd ed., 1966, p. 18).

[36] R. S. Panchamukhi, 'The Badami Rock Inscription of Vallabheśvara, Śaka 465', *Epigr. Ind.*, XXVII, 1919, pp. 4-9; D. C. Sircar, *Select Inscriptions*, 2nd ed., 1965, p. 482.—The words used are: *Śakavarṣeṣu catuśśateṣu pañcaṣaṣṭiyuteṣu*, 'In the Śaka-years four hundred, increased by sixty-five'.

[37] D. C. Sircar, *Successors of the Sātavāhanas*, Calcutta, 1939, p. 176.

[38] D. C. Sircar, *Indian Epigraphy*, 1965, p. 263. In addition, Sircar emphasizes that the Jain preference of the Śaka era is explained by the fact that, in their tradition recorded in works like the *Kālakācāryakathā*, the Śakas are represented as the defenders of the Jain faith.

Whereas the primary expansion of the Śaka era is apparently connected with Jainism, this does not apply to its secondary expansion from South India to other parts of the subcontinent and beyond. The Śaka era is apparently one of a considerable number of aspects of ancient Indian civilization which were adopted and assimilated by the 'clerks' of South East Asia who became familiar with Indian religion and erudition.[39] The precise part of the subcontinent from which the era was adopted cannot be determined, but there seems little doubt that it was from somewhere in southern India.

Methods of dating in Old Javanese, Old Malay and Old Balinese inscriptions and texts

As already indicated earlier, Indonesian inscriptions were almost invariably dated. As they were documents with important legal implications, normally dealing with the transfer of land and the conferment upon religious institutions of important rights and privileges, they had to satisfy a number of formal requirements, including precise dating. It seems that, as in modern society, no document was regarded as legally binding unless such conditions were fulfilled. For literary, religious and didactic texts there was no such emphasis on proper dating—in accordance with the practice in most other parts of the world—but it is remarkable that nonetheless many works of this kind mention their dates of composition in addition to the name and office of their authors and other details.

As far as the dating of temples and other religious monuments is concerned, it is striking that no monument earlier than the Majapahit period is itself dated, although dates were probably put in the foundation charters, whenever these were issued.

The oldest Sanskrit inscriptions of Java, those of Pūrṇavarman of Tārumā in the western part of the island, are not dated in any era, but one of these contains a date in regnal years.[40] This is a very common method in ancient India and Sri Lanka, even in relatively late times (such as in the inscriptions of the Pālas of Bengal and Bihar or the Chōḻas of the Tamiḻnāḍu), but rare in Indonesia and in other parts of South East Asia. The date, in this case, is 'the twenty-second year of [his] increasing, prosperous reign' (of Pūrṇavarman). Subsequently two precise dates are given, indicating the beginning and the completion of

[39] F. D. K. Bosch, 'The problem of the Hindu colonisation of Indonesia', *Selected Studies in Indonesian Archaeology*, 1961 (original Dutch address of 1946), pp. 3-22.

[40] See note 23 above.

the digging of a canal. These dates are: (1) the date on which the work (the digging of a canal) was begun, viz. the eighth *tithi* (lunar day) of the *kṛṣṇapakṣa* ('dark half', i.e. of the waning moon) of Phālguna (the last month of the year), and that on which it was completed, viz. the thirteenth *tithi* of the *śuklapakṣa* ('bright half') of Caitra (the first month, evidently of the following year).[41] The learned editor of the inscription pointed out that, as also the total number of days needed for the completion of the work is mentioned, viz. twenty-one full days,[42] it may be concluded that the months were counted from the new moon (*amānta*), the system that is prevalent in southern India.[43] The dating generally corresponds to that used in contemporary Pallava inscriptions. Thus the Narasarāopeṭ inscription of Siṃhavarman[44] is dated the fourth year of his victorious reign, on the fifth (*tithi*) of the bright half of Vaiśākha (*s[v]avijaya-rājya-sa[ṃ]vatsare caturtthe vaiśākha-śukla-pakṣa-pañcamyāṃ*). Also the contemporary Vākāṭaka inscriptions of the northern Deccan used similar methods of dating. Thus the Chammak (Maharashtra) Copper-plate Inscription of Pravarasena II is dated the eighteenth regnal year of this king on the thirteenth (*tithi*) of the bright half of Jyeṣṭha (Jyaiṣṭha).[45] It is interesting to compare the different method used in the Gupta inscriptions of northern India during the same period. There we normally find, in addition to the number of the year (of the Gupta era) and the name of the month the number of the solar day (*divasa*) without reference to the waxing or waning moon.[46] It seems therefore clear that Pūrṇavarman followed in his Tugu inscription the method of dating as used by the Pallavas, Vākāṭakas and other dynasties of the Deccan in or about the fifth century A.D.

The earliest known South East Asian inscription dated in the Śaka era is one in Cambodia, viz. the Khmer inscription of Aṅkor Bórĕi,

---

[41] J. Ph. Vogel, 'The Earliest Sanskrit Inscriptions', *Publ. Oudh. Dienst*, I, 1925, p. 32, line 3.

[42] Both the day of the beginning and that of the completion of the canal are evidently included in this total.

[43] D. C. Sircar, *Indian Epigraphy*, pp. 224-226.

[44] H. Krishna Sastri, *Epigr. Ind.*, XV, 1910, pp. 254 f.; D. C. Sircar, *Select Inscriptions*, 2nd ed., 1965, No. 67, pp. 469-472.

[45] J. F. Fleet, 'Inscriptions of the Early Gupta Kings', *Corp. Inscr. Ind.*, III, 1887 (reprint 1970), No. 55, pp. 235-243, 11. 60 f.; D. C. Sircar, *Select Inscriptions*, 2nd ed., 1965, No. 62, pp. 442-449; *V. V. Mirashi*, 'Inscriptions of the Vākāṭakas', *Corp. Inscr. Ind.*, V, 1963, No. 6.

[46] This is especially the case with the Gupta land grants from present Bangladesh. A typical example is the Damodarpur copper-plate inscription of the reign of Kumāra Gupta I (R. G. Basak, *Epigr. Ind.*, XV, 1910, pp. 130 f.; D. C. Sircar, *Sel. Inscr.*, pp. 290-292), line 1, first side: *samba* 100 20 4 *phālguṇa-di* 7, i.e. *saṃvat 124, phālguna-divase* 7, 'In the year 124 (of the Gupta Era = A.D. 444), the 7th day of Phālguna'. The Gupta inscriptions in verse show much more diversity in methods of dating.

dated Śaka 533 (A.D. 611).[47] This date falls 68 years after the earliest Indian example of the use of the Śaka era among the Western Cālukyas but 72 years before the earliest known Indonesian inscription dated in the Śaka era: the inscription of Kĕdukan Bukit (Palembang, South Sumatra) of Śaka 605 = A.D. 683.[48] Owing to the paucity of inscriptions in South East Asia in about this period it is not certain that the time lag between the appearance of the era in India and Cambodia, and again between its first appearance in Cambodia and Indonesia is significant, but it may not be without interest to note that it fits well into the general pattern of the spread of Indian influence in South East Asia. For it is well known that Indian influence had been established in Cambodia (Fu-nan) [49] a considerable time before it was first attested in Indonesia. This suggests, of course, by no means that elements of Indian culture should have reached Indonesia by way of Fu-nan. As far as chronology is concerned this would be quite unlikely because the methods of dating seem different in the two regions. Thus, the fairly numerous Old Khmer inscriptions of the seventh century A.D. not only mention the Śaka year, month, *pakṣa* and the ordinal number indicating the *tithi*, but also invariably add the *nakṣatra*.[50] The Indonesian inscriptions, on the other hand, rarely mention the *nakṣatra* before the second quarter of the tenth century.[51] As the *nakṣatra* is occasionally mentioned in Indian dates, at least from the time of the Kuṣāṇas,[52] it would seem that, at least as far as chronology is concerned, the Khmer inscriptions keep closer to the Indian tradition than those of Indonesia. In contrast, the Old Javanese inscriptions regularly mention the three week-days, as discussed earlier.

Although any date is precisely defined by the year, month, *pakṣa*

---

[47] Nos. K. 557 & 600, G. Cœdès, *Inscriptions du Cambodge*, VIII, 1966, pp. 166 f. and 172 f.; text and translation in *Inscr. du Cambodge*, II, 1942, pp. 21-23.

[48] G. Cœdès, 'Les inscriptions malaises de Çrīvijaya', *B.E.F.E.O.*, XXX, 1930, pp. 29-80; R. Ng. Poerbatjaraka, *Riwajat Indonesia*, I, 1952, pp. 33-35.

[49] L. Malleret, L'Archéologie du Delta du Mékong', I-IV, *Publ. E.F.E.O.*, 1959-1963.

[50] With often one or two other details. Thus, already in the oldest dated inscription of Cambodia (see note 47 above) one reads: *traitriśottarapañcaśata śaka parigra[ha] trayodaśī ket māgha puṣya-nakṣatra tulalagna*, '533 çaka, 13e jour de la lune croissante de Māgha, mansion lunaire Puṣya, la Balance étant à l'horizon'.

[51] There are, however, a few exceptions, such as the Sanskrit inscription of Dinaya (Dinoyo or Dinojo), dated A.D. 760, which mentions the nakṣatra Ardrā (*ardrārkṣe*, line 14; *Tijdschr. Bat. Gen.* LXXXI, 1941, pp. 499-51) and the Stone Pillar Inscription of Caṇḍi Asu dated A.D. 874 (Brandes, *Oud-Javaansche Oorkonden*, No. XI; Stutterheim, *Tijdschr. Bat. Gen.*, LXXXIV, 1934, pp. 86-88), as well as some inscriptions of Balitung and Dakṣa.

[52] The *nakṣatras*, which have played an important part in Indian astrology since Vedic times, are first mentioned in the dates of some inscriptions of the Kuṣāṇas and of the Western Kṣatrapas, e.g. the Gunda inscription of (Śaka) 103 (A.D. 181), mentioning the *nakṣatra* Rohiṇī (or Śravaṇā ?); see D. C. Sircar, *Select Inscr.*, p. 182, line 3.

and *tithi*, most inscriptions give more details and, especially from the tenth century, many more details are added until, by the thirteenth century the maximum is reached with dates consisting of no fewer than fifteen elements. This progressive expansion is illustrated in the adjoining chart. After the thirteenth century there is no further increase in the number of elements of the date; many later inscriptions have, in fact, less elaborate dates.

Such very elaborate methods of dating seem characteristic of Java and other parts of Indonesia where Javanese influence was strong, such as Bali and parts of Sumatra at the end of the thirteenth and during the fourteenth century. The inscriptions in Old Malay and Old Balinese, as well as the Sanskrit inscriptions found outside Java, show different and much less elaborate methods of dating, which will be discussed separately. It should also be noted that these elaborate datings are not, it seems, found in mainland South East Asia.[53] This tendency seems therefore typical of Java. The reasons for these very detailed datings seem obscure. Stutterheim's view, [54] implying that all the astronomical details are given to indicate that the foundations mentioned in the inscriptions took place at auspicious times, may seem likely but cannot be substantiated. Unlike the Indian calendars used for calculating horoscopes etc. (*pañcāṅgas*), which are based on the week-day of the seven-day week (*vāra*), the lunar mansion (*nakṣatra*), the lunar day (*tithi*), the 'conjunction' (*yoga*) and the half-*tithi* (*karaṇa*),[55] Indonesian horoscopes are mainly based on combinations of days in the three types of weeks and the name of the *wuku*,[56] in addition to the more common elements such as months and phases of the moon. Such Indian theoretical units as *nakṣatra*s, *yoga*s, *tithi*s and *karaṇa*s appear to have no place in any known Indonesian system of divination. It is, of course, not impos-

[53] It is particularly striking that the numerous inscriptions of Cambodia have relatively short dates with few, if any, details other than those indispensable for defining the precise time. In this respect there is no clear difference between the earlier and later inscriptions, which all have datings of the type illustrated in note 50 above. There are a few exceptions, viz. dates with particular astronomical details (e.g. inscription I of Tuol Komnap, Cœdès, *Inscr. du Cambodge*, III, pp. 129 f., mentioning the zodiac signs in which each of the planets stood at the time of the consecration of a Viṣṇu image), but there is no gradual increase in the number of elements of the date in the course of the centuries, as we see clearly in Java.

[54] W. F. Stutterheim, 'Een vrij overzetveer te Wanagiri (M.N.) in 903 A.D.', *Tijdschr. Bat. Gen.*, LXXIV, 1934, pp. 269-295, note 2 to p. 289.

[55] See the excellent account by P. V. Kane, *History of Dharmaśāstra*, Vol. V, Part I, Poona, 1958, especially Chapter XIX, pp. 686-718.

[56] The most complicated system, combining weeks of three, four, five, six, seven, eight and nine days, is that represented by the Balinese *tikas*, which will be discussed at the end of this article (see note 17 on p. 42). Cf. also Alfred Maass, 'Astrologische Kalender der Balinesen', *Feestb. Kon. Bat. Gen.*, II, Weltevreden 1929, pp. 126-156, and Th. Pigeaud, 'Javaansche Wichelarij en Klassificatie', *ibid.*, pp. 273-290.

sible that this was different in ancient times, but there is at least no positive indication that this was indeed the case. As long as no such evidence has come to light there is no reason to interpret the addition of more and more sophisticated elements in the dating of inscriptions as increasing emphasis on astrology. It therefore seems that a completely different explanation of the gradually increasing length of the dates in inscriptions is called for.

In different fields of ancient Indonesian civilization one can observe interesting developments. Thus Old Javanese ('Kawi') script tends to become more stylized and decorated during the course of the centuries —a development which appears to reach its climax in the thirteenth and fourteenth centuries.[57] Gradually increasing emphasis on detail and on decorative elements, sometimes to the extent of almost concealing the functional elements, have often been noticed in Javanese art and architecture. The elaboration of the dates in inscriptions may reflect a similar tendency. It should finally be noted that it is not only the date which increased in length, but also most other parts of the inscriptions. Thus, in the Pĕnanggungan Copper-Plate Inscription of Kṛtarājasa,[58] dated A.D. 1296, which provides a clear example of the maximum length of the date of fifteen elements, this elaborate date occupies not even three full lines of the 126 lines of the extant part of the inscription; as the text ends in about the middle of the inscription,[59] the elaborate date does not occupy much more than 1 or 1½% of the inscription as a whole. As a comparison I may mention the Minto Stone Inscription of Wawa (A.D. 928) where the dating occupies a little more than one line out of approximately a hundred or the Ngabean Copper-Plate Inscription of Kayuwangi (A.D. 879) with a date taking about five-sixth of one line out of a total of 30½, which is nearly 3%.[60] It therefore appears that the gradual lengthening of the date in the inscriptions reflects the general tendency of increasing the length of the inscriptions. In so doing the clerks who drafted the texts of the inscriptions probably wanted to assure that such an important part as the date retained its emphasis. With a

[57] Cf. the present author's *Indonesian Palaeography*, 1975, pp. 47-52.

[58] R. Ng. Poerbatjaraka, 'Oorkonde van Kṛtarājasa uit 1296 A.D.', *Inscr. Ned.-Indië*, I, 1940, pp. 33-49 (with introduction by W. F. Stutterheim).

[59] It ends in the middle of the list of the so-called *manilala drawya haji*, the long enumeration of all kinds of people who could claim some kind of tax from the villagers. Comparison with more or less contemporary charters shows that this is usually about the middle of the inscription. There is, however, a considerable amount of variation in the length of different sections (some inscriptions having very lengthy imprecations against evil-doers at the end, others no more than a single sentence), so one cannot be quite sure about the original length of the inscription of Kṛtarājasa.

[60] Brandes-Krom, *Oud-Javaansche Oorkonden*, 1918, No. XXXI, pp. 42-49 and No. XII, pp. 14-16.

date consisting of fifteen elements they went about as far as they could possibly go, for it is difficult to see what more could be added.

A point in favour of this view is the fact that even in the Majapahit period, when we generally find the most elaborate dates of all, there are also a number of very short charters with correspondingly short, sometimes incomplete, dates. Thus, the Charter of Rěněk (A.D. 1379), consisting of just over eighteen lines of writing, has a short date a little before the end of the text (Plate II, *verso*, line 4) occupying about half a line with somewhat cryptic abbreviations.[61] A number of other brief charters of the latter half of the fourteenth century have even shorter dates consisting only of the last figure of the year and the name or number of the month,[62] or even of nothing but the year.[63] It was apparently felt that a detailed or lengthy date would be out of proportion in a very short edict.

As to the elaborate dates there is no need for a detailed discussion here, but a few brief remarks may be useful. Although any date is fully and unequivocally defined by the first four of the fifteen elements, the additional details are of direct interest to the epigraphist and the historian in so far as they may either confirm the correctness of the reading of the essential parts of the date or help to reconstruct these essential parts in the case of partial uncertainty of the reading due to damage or other factors. The importance of such details, in particular the names of the three weekdays and of the thirty weeks (*wuku*) in the Old Javanese dates, have been studied with scrupulous precision by L.-C. Damais.[64]

[61] W. F. Stutterheim and Th. Pigeaud, 'Een Javaansche oorkonde uit den bloeitijd van Majapahit', *Djåwå*, VI, 1926, pp. 195-204; Th. Pigeaud, *Java in the Fourteenth Century*, I, 1960, pp. 118 f.—The date is given as: *tithi, wa, pang, ba, ka, 9, sirāḥ 1*, translated by Pigeaud (*op. cit.*, III, p. 170) as: 'Date: *Wage* (a day of the 5 days week), *Pahang* (the 16th week of the 30 weeks *wuku* year), Wednesday in the 9th (month, *Caitra*, March-April), head (i.e. unit of the year's number) 1'. The precise interpretation of this date, one of the few which, as far as I can see, have not been discussed by Damais, needs some correction. It can be seen from the table of the *wuku* cycle of 210 years (see L.-C. Damais, 'Études d'Épigraphie Indonésienne', IV, *B.E.F.E.O.*, XLVII, Fasc. 1, 1955, p. 253) that Wednesday of the *wuku* Pahang is always combined with Kaliwon of the five-day week. As, moreover, the inscriptions invariably enumerate the different week days in the order of the six-, the five-, and the seven-day weeks, I am sure that the abbreviations *wa, pa* and *bu* mean Wās, Paḥing and Wednesday, which fall in the second *wuku*, Landěp. In addition, the ninth month, in this period, should not be Caitra but Mārggaśira.

[62] Thus, the charter of Patapan (Pigeaud, *Java* etc., I, p. 121) gives: *tithi jyeṣṭa . sirah 7*, 'Date: (month of) Jyaiṣṭha, unit (of the year): 7', i.e. probably 1307 Śaka = A.D. 1385.

[63] One of the Biluluk charters (Pigeaud, *Java* etc., p. 115) gives merely *i śaka* 1288, i.e. A.D. 1366.

[64] L.-C. Damais, 'Études d'Épigraphie Indonésienne. I. Méthode de réduction des dates javanaises en dates européennes', *B.E.F.E.O.*, XLV, 1951, pp. 1-41; II. 'La date des inscriptions en ère de Sañjaya', *B.E.F.E.O.*, XLV, 1951, pp. 42-63; III. 'Liste des principales inscriptions datées de l'Indonésie', *B.E.F.E.O.*, XLVI, 1952, pp. 1-105; IV. 'Discussion de la date des inscriptions', *B.E.F.E.O.*, XLVII, 1955, pp. 7-290; V. 'Dates des manuscrits et documents divers', *B.E.F.E.O.*, XLIX, 1958, pp. 1-239; 'Le calendrier javanais', *Journ. Asiat.*, 1967, pp. 133-141.

Any given combination of three weekdays (of the six-, five- and seven-day week) occurs only once in 210 days; combined with one of the twelve months and one of the thirty *tithi*s, this amounts to once in 75,600 days or 260.69 years. In other words, since palaeography, proper names etc. always enable us to place an inscription within two and a half centuries, we are in a position to reconstruct the figures of a totally obliterated year if these other details have remained legible. In practice, however, these data are not always available in full. Especially the three weekdays, upon which the calculation of possibilities depends to a considerable extent, are almost always expressed by abbreviations. Thus, in the seven-day week *bu* (for *Budhawāra,* Wednesday) and *brĕ* (for *Brĕhaspatiwāra*, Thursday) do not differ greatly, nor do *śa* (for *Śanaiścarawāra*, Saturday) and *śu* (for *Śukrawāra,* Friday). Similar problems may arise for *pa* and *wa* in the five-day week and *pa*, *wa* and *wu* in the six-day week. It is obvious that uncertainty about even one of these abbreviations would greatly reduce the chances for a successful reconstruction. The other details may therefore be of some interest.

Apart from the *wuku* name, regularly mentioned from the second quarter of the tenth century and sometimes in earlier inscriptions, and of interest as a possible confirmation of the reading of the weekdays, the *nakṣatra*s and *yoga*s are of particular interest. Both are, like the *wuku*s, regularly mentioned from the second quarter of the tenth century. *Nakṣatra*s, 'lunar mansions', are based on cycles of 27.322 days, the so-called sidereal month during which the moon covers the length of the ecliptica. The cycle is, in the most common system of calculation, divided into 27 constellations of equal length, each therefore corresponding to the time in which the motion of the moon equals 13° 20′, i.e. 1.012 days.[65] There are, however, two other systems, but it seems difficult to be sure which was actually used in ancient Java, for the three systems mostly yield the same final result. As to the names of *nakṣatra*s it may be of interest to note that each also has a tutelary deity (*dewatā*), normally mentioned immediately after the name of the *nakṣatra.*[66] The name of the deity, which is really superfluous, has for the modern epigraphist the advantage that it may confirm (or disprove) his reading of the *nakṣatra* name and so, to a lesser extent, may the name of the *maṇḍala*, mentioned only in the Old Javanese inscriptions from the eleventh to the fifteenth century.

[65] W. E. van Wijk, 'On Hindu Chronology', IV, *Acta Orient.*, IV, 1925, pp. 55-80, in particular, pp. 62-64 and Table X on pp. 75-77. Cf. also by the same author: *Decimal Tables for the Reduction of Hindu Dates from the Data of the Sūrya-Siddhānta,* The Hague, 1938.

[66] W. E. van Wijk, *ibid.*, pp. 64 f. and Table XI, pp. 75-77.

As to the *yogas*, defined as 'the time during which the joint motion of sun and moon amounts to 13° 20',[67] its duration corresponds to c. .941 days. Although this is a somewhat artificial unit, it is almost invariably mentioned in the inscriptions that give the name of a *nakṣatra*. There are also 27 *yogas*, although there also exists a system of 28, which is, however, unattested in Indonesia; the length of a complete cycle equals 25.420 days.

In this connexion I may call attention to a curious expression found in a small number of inscriptions from the middle of the ninth century, *viz. hana ryy-umaḥnya*, mentioned at the end of the brief dates just after the three weekdays.[68] The expression apparently corresponds to Sanskrit *svagṛhe*, 'while (the sun or the moon) is in its own house'. Calculation of the dates in which the expression is found shows that not the lunar mansion (*nakṣatra*) is meant but the zodiac sign (*rāśi*) which is Karkaṭa, 'Cancer'.[69] Although the zodiac signs are not normally mentioned in inscriptions until much later, there are occasional references to them in some of the Sanskrit inscriptions, such as the Canggal inscription of A.D. 732 (*Kumbha-lagne* in v. I, i.e. 'Aquarius') and the same constellation in the Dinoyo inscription of A.D. 760, v. VI.[70] It is interesting to add that Kumbha, in the Canggal inscription, is said 'to be known as a stable sign' (*sthirāṅśa-vidite*), a clear reference to the classification of zodiac signs in some Indian astrological texts, such as the *Bṛhaj-jātaka*, as well as in Ptolemy's *Tetrabiblos*.[71] A synonym of *sthira*, viz. *dhruva* is used in the Dinoyo inscription.

As to the numerous other elements of the dates, no detailed discussion is necessary. Many inscriptions mention the place of the 'planet' (*grahacāra*), presumably Jupiter at sunrise,[72] as well as the *maṇḍala*,

[67] P. V. Kane, *History of Dharmaśāstra*, Vol. V., Part I, 1958, pp. 501-504.

[68] The inscription of Caṇḍi Perot, *Oud-Jav. Oork.*, V & VI, *Pras. Indon.*, II, 1956, No. IX, pp. 211-243, line 4 and line 3 respectively.; Inscription of Argapura, *Oud-Jav. Oork.*, No. VIII, p. 10. For *svagṛhe* cf. P. V. Kane, *op. cit.*, p. 576: 'Certain rāśis are declared to be the houses (*svagṛha*) of planets . . . . . The Sun and the Moon have each only one rāśi as *svagṛha*. . . . .'.

[69] The mention of the *rāśi* became a regular part of the date only in the later part of the Kaḍiri period, after the reign of Jayabhaya, but it is occasionally found in earlier inscriptions of the Kaḍiri period, such as in the stone inscription of Sirahkĕting (Ponorogo) of A.D. 1104, *Oud-Jav. Oork.*, LXVI (p. 149), A (front), line 7: *wṛścikarāśi*, 'zodiac sign: the Scorpion', which is consistent with the month of Kārttika in which the inscription is dated. Brandes wrongly read: *mūṣika* instead of *wṛścika*.

[70] Line 16 of the Dinoyo inscription.

[71] P. V. Kane, *op. cit.*, p. 567: the four zodiac signs of *vṛṣabha* (bull), *siṃha* (lion), *vṛścika* (scorpion) and *kumbha* (aquarius) are regarded as *sthira* (fixed or lasting).

[72] One of the earliest examples is the Singosari inscription of A.D. 891, *Oud-Jav. Oork.*, XIX, front, line 3 (*pūrbwasthāna*). The place of the planet is occasionally mentioned in inscriptions of Balitung and Siṇḍok but becomes a regular part of the date only from the time of Airlangga.

each of the eight parts of the sky where the *nakṣatra* is placed at sunrise.[73] The *karaṇa*, regularly mentioned from the time of Airlangga but occasionally also much earlier,[74] is one-half of a lunar day (*tithi*), corresponding to c. .492 ordinary days.[75] The naming of *karaṇas* is curious: they go in four cycles of seven (*viṣṭi, bava, vālava, kaulava, taitila, gara* and *vaṇija*), but the first *karaṇa* of each month is not called *viṣṭi* but *kiṃstughna*, while the three last *karaṇa*s of the month are called *śakuni, nāga* and *catuṣpada*.[76] For the epigraphist the main interest of the *karaṇa* lies in the fact that it enables him to check the correctness of his reading of the *tithi*. On the other hand, the *karaṇa* is, together with the *tithi, vāra* (day of the seven-day week), *nakṣatra* and *yoga*, one of the elements of the Indian *pañcāṅga*, the astrological calendar.[77] As already argued earlier, there is no evidence that this *pañcāṅga* system was ever used in Indonesia, its place largely being taken by the three weekdays and the *wuku*.

At this stage it may be useful to summarize the special features of the Old Javanese system. As in many parts of India, especially the Deccan, the *amānta* system is followed in which the months are counted from one full moon to the next. The intercalation of months seems to follow the rules as laid down in e.g. the *Sūryasiddhānta*, although it is difficult to ascertain whether the corrections (*bīja*), as prescribed in the more sophisticated systems (such as the *Brahmasiddhānta* by Brahmagupta) were ever applied. There is similar uncertainty concerning the precise system of *nakṣatra*s, as there were three systems in current use in ancient India: the equal space system with the sidereal month divided into 27 equal parts, that of Garga, which took the different lengths of the con-

[73] The sections of the sky are always mentioned by the name of the tutelary deity of that particular section, such as *mahendra-maṇḍala*, the East, *āgneya-maṇḍala*, the South-East, *baruṇa* or *bāruṇya-maṇḍala*, the West, and *bāyabya-maṇḍala*, the North-East. The *maṇḍala* is usually, but by no means always, mentioned in inscriptions from the Kaḍiri period. It is curious that the names of the deities are in this context invariably spelt with *ba* instead of *va* (*wa*), which may suggest a source from north-eastern India, perhaps Bengal, for these *maṇḍalas*. I am, however, unaware of any inscriptions from Bengal (or from Bihar, Assam or Orissa) which mention the name of the *maṇḍala* in their dates.

[74] The *karaṇa* (often spelt *kāraṇa* in Old Javanese inscriptions) is, unlike the place of the planet and the *maṇḍala*, an important element of ancient Indian astrology and chronology (see note 77 below). The earliest known reference in Indonesia is in the Ponorogo copper-plate inscription of Balitung, A.D. 901 (*Oud-Jav. Oork.*, No. XXIII, p. 28); after that it is only occasionally mentioned in other tenth-century inscriptions, but becomes a regular part of the dates in inscriptions from the time of Airlangga in the second quarter of the eleventh century.

[75] W. E. van Wijk, *Acta Orient.*, IV, 1925, pp. 58 f. and Table IX on pp. 73 f.; *Decimal Tables* etc., Table IV-B.

[76] Most of these names are actually found in Old Javanese inscriptions. About half of the names are those of animals or classes of animals, some others are obscure; cf. P. V. Kane, *op. cit.*, pp. 78 f.

[77] P. V. Kane, *op. cit.*, pp. 707-718.

stellations into account, and that of Brahmagupta with 28 *nakṣatra*s and, but for this additional *nakṣatra* (Abhijit) of only 0.322 day, running in round numbers of ½, 1 and 1½ days.[78]

The most striking feature of the Old Javanese system is, however, its consistency. Apart from two inscriptions in the Sañjaya era,[79] which use exactly the same system of chronology but with a different starting point, viz. A.D. 717, presumably the date of the accession or consecration of king Sañjaya, and, at least before c. A.D. 1500, a very small number of inscriptions in the Hijrah era, all the inscriptions, literary texts, dates on monuments, including even many dates on Muslim tombs, are in the Śaka era of A.D. 78. This consistency stands in clear contrast to the diversity of eras and the use of regnal years in ancient India.

As the inscriptions of Sumatra and Bali are less numerous than those of Java it is not possible to get a clear impression of the system of chronology followed in them. The inscriptions of Sumatra generally seem to use similar systems of dating as those of Java, though without the typically Javanese elements such as the *wuku*s and the days of the five- and six-day weeks. There are, however, two interesting exceptions. The first is the inscription of Hujung Langit, near Liwah, Regency of Krui, Bengkahulu ('Bencoolen'), which mentions not only the name of a *wuku* (Kuniṅan) but, it seems, also the days of the five- and six-day weeks.[80] Of the Śaka year only the first figure (9) has remained legible, but Damais has reconstructed the figures of the year as (Śaka) 919, i.e. A.D. 997. Although the basis on which the calculation is founded is somewhat uncertain, it cannot be denied that the use of typically Javanese elements in the date—in an otherwise apparently Old Malay inscription—may indicate influence from Java. If the reconstruction by Damais of the date is correct this would not be surprising as the Javanese expedition of 991/992 may well have led to the establishment of Javanese settlements in parts of southern Sumatra.

The only other Sumatranese inscription using the *pañcavāra* and *ṣaḍvāra* is that inscribed on the basis of the Amoghapāśa statue of

[78] For these three systems of counting *nakṣatra*s see W. E. van Wijk, *Acta Orient.*, IV, 1925, pp. 62-64, as well as table X, First and Second Parts, pp. 75-77. I have not come across any example of Abhijit, which might plead against the system of Brahmagupta. It should, however, be admitted that the absence of Abhijit may not be statistically significant in view of the considerable number of *nakṣatras* (28) in relation to the fifty odd inscriptions mentioning a *nakṣatra*. If one also considers the brevity of the *nakṣatra* (.322 day) it follows that there is hardly a one in two chance for it to occur, even if the 28 nakṣatra system were in use.

[79] L.-C. Damais, 'Études d'Épigraphie Indonésienne', II, La date des inscriptions en ère de Sañjaya', *B.E.F.E.O.*, XLV, Fasc. 1, 1951, pp. 42-63.

[80] L.-C. Damais, 'Études Soumatranaises', I & II, *B.E.F.E.O.*, L, Fasc. 2, 1962, pp. 275-310, in particular pp. 285 ff.

Padang Roco' (Sungai Langsat) of A.D. 1286.[81] Apart from the absence of the *karaṇa*, given in most later Javanese inscriptions from the eleventh century and occasionally earlier, the dating of the Amoghapāśa contains the maximum number of elements, as in the inscriptions of the Singhasari-Majapahit period in general. As, in addition, it follows from the contents of the inscription on the pedestal that it was actually inscribed in East Java, from where it was shipped to South Sumatra, it can be hardly surprising that the method of dating corresponds to Javanese, not Sumatranese, standards.

As to the inscriptions from Bali there are some interesting differences from those of Java. The oldest inscriptions, from A.D. 882 to 942 are all entirely in Old Balinese, except for the pillar of Sanur, most of which is in Sanskrit. They start with the formula *yumu pakatahu sarbwa*, 'You all should know (that)', whereas the date is placed at the end in a concise form. It consists of the month, the *pakṣa*, the *tithi*, the weekday of the three-day week and, finally, the Śaka year.[82] The subsequent group of inscriptions, from A.D. 951 to 983, though still in Old Balinese, place the date at the beginning, while the words *yumu pakatahu sarbwa* are no longer found. The inscriptions then begin with *i(ng) śaka*, followed by the figures of the year, the name of the month (preceded by the word *bulan*), the *pakṣa*, the number of the *tithi* (as always expressed by a Sanskrit ordinal in feminine gender, except for the first day, *pratipāda*), and, finally, the day of the three-day market week (preceded by the words *rggas pasar*, 'at the time of the market day').[83] This three-day week, based apparently on the place (in or near the capital?) where the market was held, is not found elsewhere and is therefore likely to be typically Balinese. From 994, however, the day of the three-day week is, except in an inscription of 1001 and one of 1016, replaced by the three Javanese weekdays, usually given in their common abbreviations, and the name of the *wuku*.[84] It is striking that this relatively sober, but absolutely unambiguous, form of dating remained the regular one in all the later inscriptions written, with but a few exceptions, not in Old Balinese but in Old Javanese. Although, in general, the inscriptions reflect strong Javanese influence in the formulation of privileges, the curse directed against evil-doers etc., the very elaborate methods of dating were not adopted by the Balinese.

---

[81] N. J. Krom, 'Een Sumatraansche inscriptie van koning Kṛtanagara', *Med. Kon. Akad. Wet.*, N. R., 1916, pp. 1-34.

[82] R. Goris, *Prasasti Bali*, I, 1954, Nos. 001-110, pp. 53-72.

[83] *Ibid.*, Nos. 201-210, pp. 72-79.

[84] *Ibid.*, Nos. 303-357, pp. 83-107.

Finally, the inscriptions of Sunda (West Java) are too few in number to give us an idea about the methods of dating used in this part of the island. After the above mentioned inscriptions of Pūrṇavarman [85] the next dated inscription [86] is that of king Jayabhūpati, incised in four stones. The inscription is in Old Javanese, and its dating also agrees with the Javanese method. It consists of the year (Śaka 952 = A.D. 1030), the month, *pakṣa*, *tithi*, the three weekdays and, finally, the *wuku*. As in the inscriptions from Bali, no astronomical details (*nakṣatra*, *yoga* etc.) are added. As to the Old Sundanese inscriptions of the Majapahit period, only one is dated, *viz.* the well known stone of Batu Tulis near Bogor, West Java.[87] The date consists, however, of the year only, which is expressed by a *candra-sĕngkala* and will be discussed in the next section.

In conclusion it appears that the inscriptions outside central and eastern Java, in particular those from Sumatra, Bali and western Java, initially used independent methods of dating but were, from the end of the tenth century, influenced by East Java, especially in the adoption of the three weekdays and the *wuku*. The influence of East Java remained, however, limited to these details and did not include the astronomical details, which were not indispensable for the determination of the precise date.

Dates in texts, on monuments and other archaeological materials

It is well known that in ancient South and South East Asia many texts, in particular those on religion and associated fields, were considered to belong to all times. The actual time at which they were composed or revealed was therefore considered immaterial. Yet, in this respect as well as in others, one can distinguish increasing emphasis on dating, especially differences in attitude as between the earlier and later periods. In India before and during the Gupta period no text is dated. The date can, in such circumstances only be determined by indirect data, and it therefore is almost always conjectural. Even when texts are not anonymous there is often so much uncertainty about the time when the author lived that his name is of little help in dating the

[85] See note 23 above.

[86] C. M. Pleyte, *Tijdschr. Bat. Gen.*, LVII, 1916, pp. 201-218.

[87] Owing to its conspicuous location a few miles south of Bogor, the former residence of the Governors-General of the 'Netherlands East Indies', as well as its connexion with the foundation of Pajajaran, the stone of Batu Tulis (the name of the village means, in fact, 'Inscribed Stone') this inscription drew the attention of very early scholars; see N. J. Krom, *Hindoe-Javaansche Geschiedenis*, 2nd ed., 1931, pp. 404 f. and the literature quoted there.

text. From about the seventh century A.D., however, the texts and their authors are usually more or less accurately fixed in time. Thus, we know that Bāṇa (or Bāṇabhaṭṭa), the author of the *Harṣacarita* and *Kadambarī*, was a contemporary of king Harṣa (or Harṣavardhana) of Kanauj (c. 606-647).[88] In Indonesia the oldest known text which, at least in one of its manuscripts, gives an indication of its date is the *Sang Hyang Kamahāyānikan*, dated during the reign of king Siṇḍok (c. 929-947).[89] About a century later the *Arjunawiwāha* explains that its author, Mpu Kāṇwa, wrote the text during the campaigns of king Airlangga (c. 1016-1042).[90] From then on the literary texts of the *kĕkawin* type [91] regularly mention the name of the king during whose reign they were composed. In many cases they even give us a precise date when the work was composed. The oldest example of this kind is the *Bhārata-yuddha* by Mpu Sĕḍah and Mpu Panuluh, completed in A.D. 1157.[92] The oldest dated prose text is, however, considerably older; it is the Old Javanese version of the *Wirāṭaparwa*, which gives a precise date, which I reproduce here in Zoetmulder's translation: 'This is as I remember it: it was the 15th of the dark moon in the month Asuji; the day was *tungle, kaliwon*, Wednesday, in the *wuku* Pahang of the year 918 of the Śaka-era. It is now *mawulu, wage*, Thursday in the *wuku* Maḍangkungan, the 14th of the dark moon in the month Kārttika. So the time is just one day short of one month.' The two dates indicated here are those of the beginning and the end of the recitation of the text in the presence of king Dharmawangsa T(ĕ)guh Anantawikrama (c. A.D. 991-1016). The two dates mentioned in the text correspond to the 14th of October and the 12th of November, 996, respectivelly.[93]

---

[88] The *Harṣacarita* actually begins in Chapter I by relating the genealogy and the youth of its author and subsequently, in Chapter II, describes how he was introduced to Harṣa's court. See E. B. Cowell and F. W. Thomas, *The Harṣa-carita of Bāṇa*, reprint Delhi-Varanasi-Patna, 1961, pp. 4-69; U. N. Ghoshal, *Studies in Indian History and Culture*, Bombay 1965, pp. 49-100; V. S. Pathak, *Ancient Historians of India, A Study in Historical Biographies*, Bombay etc., 1966, pp. 30-55.

[89] J. Kats, *Sang Hyang Kamahāyānikan*, The Hague, 1910, p. 118. The passage mentioning king Siṇḍok (A.D. 929-947) as well as the name of the author Śri Sambhārasūryāwaraṇa in part of an introduction found only in MS *C*.

[90] N. J. Krom, *Hindoe-Javaansche Geschiedenis*, 2nd ed., 1931, p. 269; C. C. Berg, 'De Arjunawiwāha, Er-Langga's Levensloop en Bruiloftslied', *Bijdr. Kon. Inst.* 97, 1939, pp. 19-94; P. J. Zoetmulder, *Kalangwan, A Survey of Old Javanese Literature*, The Hague 1974, pp. 234-249.

[91] *Kĕkawin* or *kakawin* indicates one of the two main types of Old Javanese poetical works, viz. that in which Indian metres and Sanskrit prosody are utilised—in contrast to *kidungs*, which are based on Indonesian metres and Indonesian prosody. See Zoetmulder, *op. cit.*, pp. 101-121.

[92] J. G. H. Gunning, *Bhārata-yuddha, Oudjavaansch Heldendicht*, 's-Gravenhage, 1903. The year is given in a chronogram expressing the (Śaka) year 1079, i.e. A.D. 1157. Cf. Zoetmulder, *op. cit.*, p. 269.

[93] Zoetmulder, pp. 95 f., who convincingly argued that the nature of the dating is a guarantee for its authenticity, which has sometimes been doubted.

The authenticity of this dating has been questioned, but without real justification, for it is difficult to see why and especially how such elaborate dates would and could have been made up in later times. What is, however, interesting is that the dating is far less elaborate than that found in contemporary inscriptions which invariably add such details as *nakṣatra* and *yoga*, and often more.[94] This simpler form of dating rather belongs to the ninth century, but then without the mention of the *wuku*, which first appears in the second quarter of the tenth century. The literary texts apparently follow a slightly different, in part more archaic, tradition in dating.

In poetry—and this includes those inscriptions that are composed in metre—it is impossible to use numerals written in figures. It is, of course, possible to write the numerals out in full but, as there is but little flexibility in this case,[95] it is difficult to fit in such, usually lengthy, words with the metre. The same problem already existed in India in metrical inscriptions. There this problem was solved by replacing the numerals by words the meaning of which suggested a certain number. Thus, one could use 'eyes' for 2, 'fire' for 3, 'Veda' for 4, etc. From early times this method was used in, for instance, astronomical works, which were often written in verse. Many conventional substitutes were established in this manner, so that the author of astronomical, mathematical etc. works normally had considerable choice for each common numeral and could thus select those that could be fitted in with the metrical pattern that was used.[96] Apart from the obvious need for such words in metrical texts, this usage had the further advantage of lessening the difficulty of memorizing the texts.

This practice spread to South East Asia. In Java, the earliest example is found in the Canggal inscription, where the date is expressed by the

[94] Unfortunately, the contemporary inscription of Sĕndang Kamal, one of the longest Old Javanese inscriptions, but, except for the first 12 lines, still unpublished (Brandes, *Oud-Javaansche Oorkonden*, LVII), has serious lacunae in the date. About thirty years later, the inscriptions of Airlangga, dated from A.D. 1021 (*Oud-Jav. Oork.*, LVIII), give very elaborate dates, which include the position of the planet, the name of the divine protector of the *nakṣatra*, the *karaṇa* and the *maṇḍala*.

[95] This may not be a strong argument as there are a considerable number of Indian inscriptions in verse using ordinary numerals. Thus, the Mandasor inscription of Kumāra Gupta and Bandhuvarman (verses 34 and 39) contains two dates (J. F. Fleet, 'Inscriptions of the Early Guptas', *Corp. Inscr. Ind.*, III, 1887, reprint Varanasi 1970, pp. 79-88), the Junagarh rock inscription of Skanda Gupta, *ibid.*, pp. 56-65) even three (verses 27, 33 and 45), and plenty of other examples can be found. In Indonesia the Sanskrit inscription of Kalasan, verse 6 (F. D. K. Bosch, 'De inscriptie van Kĕloerak', *Tijdschr. Bat. Gen.*, LXVIII, 1928, pp. 1-64, in particular p. 59), also uses numerals in verse. It is perhaps not without interest to add that in the last mentioned case, as well as in several of the Indian inscriptions, the *Āryā* metre is used, which, unlike the metres used in *kĕkawins*, admits of considerable flexibility.

[96] Cf. J. Filliozat in Renou-Filliozat, *L'Inde Classique*, II, 1953, p. 182.

words *śruti*, 'sacred text' and therefore a synonym of Veda, i.e. 4, *indriya*, 'senses', i.e. 5 and *rasa*, 'sensations', i.e. 6. Contrary to the order in which numerals are written, the symbolic words always start with the units, followed by the tens, then the hundreds and, wherever required, the thousands. The words just mentioned therefore represent 654 of the Śaka era, corresponding to A.D. 732.[97] This example is from a Sanskrit inscription. The oldest example where the same method is used in an Old Javanese inscription is of more than a century later, when an inscription of uncertain origin but probably from, or from the neighbourhood of, the Ratubaka plateau uses the words *wualung*, '8', *gunung*, 'mountain', i.e. '7' and *sang wiku*, a synonym of *ṛṣi*, i.e. '7', together representing (Śaka) 778 or A.D. 856.[98]

Apart from the metrical inscriptions the use of symbolic words to replace figures spread to other texts, mainly, it seems, for mnemotechnical reasons. This is most striking in the *Pararaton*, an Old Javanese prose text which, in its present form, must be dated after the end of the fifteenth century.[99] Krom's analysis has clearly shown that the text has grown out of a list of dates followed by brief indications of events taking place at such dates. Subsequently numerous narratives of varying lengths have been inserted wherever such narratives were available. The actual 'skeleton' of the text is therefore undoubtedly old, and the use of dates is likely to reflect the practice of the early Majapahit period.

In the *Pararaton* the words with numerical value have sometimes been chosen in such a manner that they allude to the event described in the text. Thus, the death of a certain Tañca, who had killed king Jayanagara, but fell himself by the revenging hand of Gajah Mada, is said to have taken place in the Śaka year suggested by 'whoever raises his hand against the king will be reduced to ashes' (*bhasmi-bhuta naṅani ratu*, i.e. Śaka 1250 or A.D. 1328).[100] There are probably many other examples of this kind but lack of precise knowledge of the circumstances under which some of these events occurred often prevents us from understanding the allusions.

---

[97] H. Kern, 'De Sanskrit-inscriptie van Canggal (Kĕḍu), uit 654 Çāka', *Verspr. Geschr.*, VII, 1917 (original article of 1885), pp. 115-128, verse 1; B. Ch. Chhabra, *Expansion of Indo-Aryan Culture*, reprint 1965, p. 45, reports the precise date and time: 'the 6th of October, 732 A.D., at one o'clock in the afternoon.'

[98] Brandes, *Catalogus Groeneveldt*, 1887, p. 382; J. G. de Casparis, 'A metrical Old Javanese inscription dated 856 A.D.', *Pras. Indon.*, II, 1956, pp. 280-330, verse 24.

[99] J. L. A. Brandes, *Pararaton (Ken Arok), of het Boek der Koningen van Tumapĕl en van Majapahit*, 2nd ed. by the care of N. J. Krom, with the co-operation of J. C. G. Jonker, H. Kraemer and R. Ng. Poerbatjaraka (= *Verh. Kon. Bat. Gen.*, LXII), 1920, especially pp. 5* and 6*; N. J. Krom, 'De samenstelling van de Pararaton', *Tijdschr. Bat. Gen.*, LX, 1921, pp. 86-103.

[100] Brandes-Krom, *Pararaton*, text, p. 34, lines 24 f.; translation, p. 129.

The *Nāgarakrĕtāgama*, dated A.D. 1365, also contains numerous dates, all expressed by *candra-sĕngkalas*. There are forty of such dates in all, not counting the date of the second colophon of the 1740 manuscript. For a complete list with their value in numbers and the corresponding dates in the Christian era the reader is referred to volume V of Pigeaud's monumental work on *Java in the Fourteenth Century*.[101] The words used in these chronograms are almost all of Sanskrit origin and contain no allusions to the events in connexion with which they are mentioned. An exception to both these statements is the earliest date after the beginning of the Kaliyuga mentioned in the text, viz. Śaka 124, expressed as *samudra naṅgung bhumi*, 'the ocean carries the earth'. This date is given as that on which the island of Madura, originally one with the land of Java (*tuṅgal mwang yawaḍāraṇi*), became separated from the main island by a narrow strait.[102] The chronogram, though not exactly mentioning this 'event', nonetheless alludes to it by suggesting the power of the ocean. Most of these *sĕngkalas* are unambiguous, but a few may raise doubt as to their numerical value.[103]

The use of *sĕngkalas* is not limited to the earlier period but extends to the present day in Java and Bali. In addition to the use of words with numerical value we find, perhaps already from the eleventh century,[104] representations which suggest certain numbers. The earliest known example is a carving on a piece of stone from Bĕlahan on the eastern slope of the Gunung Pĕnangungan. It shows the head of Rāhu, wearing the headdress of an ascetic, biting in the sun. As Rāhu is one of the nine 'planets' (*navagraha*) and ascetic, regarded as a synonym of *ṛṣi* (*rĕsi*), suggests 7, while the sun is normally 1, this would yield the figures 9, 7 and 1. As Bĕlahan is thought to be the place where the remains of king Airlangga were deposited, Stutterheim suggested that the carving gives the date of Śaka 971, corresponding to A.D. 1049, presumably the date of the death of Airlangga. This date, following the date of Airlangga's last known inscription of A.D. 1042, seems quite plausible, but this does not necessarily imply that the interpretation of the carving is correct. If Rāhu is biting the moon, the value of the chronogram should be 1379 ('Rāhu, ascetic, is biting the moon') but if he is merely seizing it with his two hands it would be 1279.[105]

[101] Th. Pigeaud, *Java in the 14th Century*, V, 1963, pp. 20-22.

[102] Canto XV, v. 2.

[103] It seems, however, that most, if not all, of such uncertainties have gradually been clarified, as will appear from Pigeaud's standard work.

[104] W. F. Stutterheim, 'Oudheidkundige Aanteekeningen, XLII. Is 1049 het sterfjaar van Erlangga?', *Bijdr. Kon. Inst.*, 92, 1935, pp. 196-202 and Fig. 5).

[105] The problem, in this particular case, is that normally such pictorial chronograms show some

In the Majapahit period such chronograms in 'pictures' are not uncommon, and this practice, mainly on buildings, has continued till recent times.[106]

Dates on temples and other religious buildings are completely absent in the earlier part of the period. No examples are known from the Central Javanese period, nor are there any clear examples before the Majapahit period. There is a possible exception, viz. the three figures 899 hewn in the rock behind the *tīrtha* of Jalatunda on the western slope of the Gunung Pĕnanggungan.[107] It is difficult to see what these figures could represent except to indicate a date in the Śaka era, which would then correspond to A.D. 977.

In the Majapahit period, on the other hand, such dates on temples and other sacred buildings are very common. Thus, a number of different dates are found on buildings of the complex of Panataran, viz. 1241 (A.D. 1319/20), 1242 (A.D. 1320/21), 1245 (A.D. 1323/24), 1269 (A.D. 1347/48), 1295 (A.D. 1373/74), 1297 (A.D. 1375/76), and 1301 (A.D 1379/80). Two of these dates were found on loose stones, the original position of which is not known, but all the others are on conspicuous spots on temples, terraces, and gateways. These dates are of great value for the architectural history of the Panataran complex.[108] The last known date is 1376 (A.D. 1454/55).[109] Even many more dates have been recovered from the more than seventy known archaeological sites on the Gunung Pĕnanggungan and the neighbouring Gunung Arjuno. Thirty-nine dates have been published.[110] If one excludes the above-mentioned date of Jalatunda, the two inscriptions of Jĕdung and two uncertain readings, all belong to the Majapahit period during the fourteenth and fifteenth century.[111] The two most recent dates are 1422 (A.D. 1500/01) and 1433 (A.D. 1511/12).[112] There are also a few pictorial chronograms.

kind of action, which determines the order in which the suggested figure values must be read, not a mere juxtaposition of words suggesting some numerical value.

[106] Especially on various buildings in the kratons of Surakarta and Jogjakarta.

[107] W. F. Stutterheim, 'Oudheidkundige Aanteekeningen, I. Gĕmpĕng', *Bijdr. Kon. Inst.* 85, 1929, pp. 479-482; F. D. K. Bosch, 'The OldJavanese Bathing-Place Jalatuṇḍa', *Selected Studies in Indonesian Archaeology*, 1961 (original article of 1945), pp. 50 ff.

[108] Krom, *Inleiding*, II, pp. 245-284; *Hindoe-Javaansche Geschiedenis,* 2nd ed., 1931, p. 382, 391, 401 etc.; M. E. Lulius van Goor, 'Notice sur les Ruines de Panataran', *Études Asiat. E.F.E.O.*, 1925, I, pp. 375-384.

[109] Krom, *Inleiding*, p. 247.

[110] V. R. van Romondt, *Peninggalan-peninggalan Purbakala di Gunung Penanggungan*, pp. 1-56, especially Lampiran E on p. 52.

[111] There are in all eight fourteenth-century dates, but no fewer than twenty-six belong to the fifteenth century or the very beginning of the sixteenth.

[112] The first of these two dates is, however, uncertain. Thus, van Romondt (*op. cit.*, p. 26) wrote: 'Above Antiquity No. II on the northern slope of the Penanggungan there is supposed to be another ruin, which was given No. XXVI. According to Mr. Gall there is at that site a date which

That of Antiquity No. LII shows a Nāga with ascetic headdress biting his tail, corresponding to Śaka 1378 (A.D. 1456/7). Many other dates on temples, other buildings, and images are known from the Majapahit period, and it is not necessary to provide a full list. Almost all, excluding, of course, the pictorial chronograms, are in bold Old-Javanese numerals protruding in relief out of a rectangular recess in one of the stones above the main entrance of a temple or at the back of a statue.[113]

The consistency with which individual buildings are dated during the Majapahit period clearly shows the importance attached to their dates. This contrasts with the earlier periods when no dates are found on the monuments but, in many cases, small votive inscriptions stating that some building was the pious donation (*anumoda*) by a certain person. The many possible implications of this difference need not concern us here, but it would seem, in general terms, that, whereas the earlier monuments were seen as timeless, there was a keen awareness of chronology in the Majapahit period.

In this connexion I may briefly mention the so-called *prasen*, 'zodiac beakers' [114] from East Java and Bali. They are bronze vessels of a particular shape (relatively wide in relation to their height) with an eight-pointed solar disc at the bottom, twelve symbols representing each of the zodiac signs (*rāśi*) running round the lower half of the body of the beaker with as many representations of divinities (?) around its upper half.[115] In addition, a number of these beakers have a four-figure date which, as the dates on the buildings, statues etc. of the Majapahit period, is contained within a rectangular frame on top of one of the 'divinities' (in this case, a bird) which for this reason is depicted somewhat below the level of the other figures. All the dated *prasen*s belong to the Majapahit period of the fourteenth and fifteenth century.[116] The tradition of making

---

can be read as 1422 i.e. A.D. 1500. Nothing more is known about this site'. (freely translated from Dutch). I have been unable to see the estampage mentioned by van Romondt.—The second of these two dates (1433, i.e. A.D. 1511) is certain.

[113] Van Romondt, *op. cit.*, fig. 37. The figure represents a Nāga (8) ascetic (7) biting (3) its tail (1).

[114] These 'zodiac beakers' caught the attention of some of the pioneers of ancient Javanese studies, notably of Friederich and Millies. The oldest study which is still of great value to us is that by J. H. F. Kohlbrugge, 'De heilige bekers der Tenggereezen. Zodiakbekers', *Tijdschr. Bat. Gen.*, XXXIX, 1897, pp. 129-141, with 29 illustrations; N. J. Krom, *Inleiding tot de Hindoe-Javaansche Kunst*, 2nd ed., 1923, p. 450; W. F. Stutterheim, *Cultuurgeschiedenis van Java in Beeld*, 1926, Figs. 151 and 152 (on page 108).

[115] The representation of the different signs of the zodiac is interesting, but cannot be discussed here. As to the divinities, if this is the correct way to describe the curious figures (one of which is styled Bhaṭāra Guru in de Tĕnggĕr area, but another apparently represents a crow), no convincing explanation has as yet been offered.

[116] The largest collection of these zodiac beakers is in the Musium Pusat at Jakarta; see W. P. Groeneveldt, *Catalogus der Archaeologische Verzameling van het Bataviaasch Genootschap*, Weltevreden, 1889, Nos. 795-802, pp. 229-231. The known dates range from A.D. 1321 to 1430; cf.

these beakers extends to relatively recent times in the Těnggěr area on the slopes of the Gunung Bromo, where they have been, or perhaps are still being, used for libations during religious, non-Muslim, ceremonies.[117] These more recent *prasens* bear no date but show another eight-pointed solar disc.

The use made of these *prasens* in recent times does not necessarily provide a clue as to their function during the Majapahit period. The presence of the zodiac signs as well as the solar disc may suggest some relation with the measurement of time, e.g. their use in religious ceremonies connected with the *rāśis* or the *saṃkrāntis*, 'the times on which the sun enters each of the twelve signs of the zodiac'. It is well known that the *saṃkrāntis* were, and still are, widely celebrated all over India.

The Muslim tombs of Trålåyå near Trowulan,[118] the site of ancient Majapahit, all have Arabic inscriptions, usually quotations from the Qur'ān, as well as dates, expressed only once in the Hijrah era but in all other cases in the Śaka era.[119] The numerals are, except for the one inscription in the Hijrah era, all written in Old Javanese script of essentially the same type or types as the numerals of the Pěnanggungan; in addition, there is one date in the Śaka era, though in Arabic script. All the dates range from Śaka 1278 (A.D. 1356/57) to Śaka 1533 (A.D. 1611/12).[120] There once were also a few older inscriptions (from Śaka 1203 = A.D. 1281/82) but these are now lost.[121] As only one of the tombs

Jan Fontein, R. Soekmono and Satyawati Suleiman, *Ancient Indonesian Art of the Central and Eastern Javanese Periods*, The Asia Society, New York, 1971, No. 87, p. 156, where it is added that 'the symbols of the Javanese zodiac differ in certain respects from those current in the Western world. The Gemini twins are replaced by paired crabs (*mithuna*) and Capricorn by a shrimp. A fish-elephant (cf. No. 102) often represents Pisces. Each sign is accompanied by a representation of a figure resembling the *Pånåkawans* or Javanese clowns, who are thought by some to represent the ancestors. Inside, on the bottom, is a tortoise. The elliptical shape of its shell symbolizes the orbit of the planets.—Despite this lucid and concise description many problems still remain and a detailed study of these *prasens* is desirable.

117 For a detailed account of these ceremonies see P. V. Kane, *History of Dharmaśāstra*, Vol. V, Part I, Poona, 1958, pp. 211-236, especially *Makarasaṃkrānti*, which is deified as a form of Durgā.

118 L.-C. Damais, 'Études Javanaises. I. Les Tombes Musulmanes datées de Trålåyå', *B.E.F.E.O.*, XLVIII, Fasc. 2, 1957, pp. 353-415, Plates XV-XXXIV. This article lists all the earlier publications on this site.

119 This is the date of Tomb No. IX, although the actual tomb stone (*maesan*) now stands on a wall in the immediate vicinity (Damais, p. 408). The date is A.H. 874 written in Arabic numerals. The inscription differs from all the others also in the fact that it mentions the name of the deceased, a certain Zaynuddīn. As Damais rightly pointed out (pp. 408 f.) it is impossible to know whether the deceased was an Indonesian or a foreigner.

120 See the list in Damais' article, p. 414.

121 In the list mentioned in the preceding note the first date refers to the planting of a Bodhi tree; the date and the brief inscription were read by Brandes, but have since disappeared; the second, third and fourth of the list have also disappeared, but the dates had been read by Verbeek. Owing to the numerous mistakes made in reading the Old Javanese numerals before the beginning of this century there is room for doubt about the correctness of the four oldest dates at Trålåyå.

actually bears the name of the deceased we here see the curious situation that the date of the tomb is regarded as more important for posterity than the identity of the deceased. The most important conclusion to be drawn from the analysis of these tombs is, however, the fact that the Śaka era was apparently so firmly established as the normal system of chronology that Islam did not, at least not at this stage in East Java, displace the older system which remained in use in central and eastern Java till somewhat before the middle of the seventeenth century.

Outside Java, however, and occasionally in Java, too, dates in the Muslim era appeared well before the end of the Majapahit period. The era itself, starting from the 15th of July, 622 A.D., the date of the Hijrah of the Prophet Muḥammad,[122] is reckoned in the same manner as in other Muslim countries and does not require any discussion here. It may, however, be useful to mention the earliest examples of the use of the era in Indonesia.

The much discussed inscription on the tomb stone of Leran (*kawĕdanan* and *kabupaten* Grĕsik, *karesidenan* Surabaya, East Java), probably dated A.D. 1082, is the oldest Muslim inscription in Indonesia. Its script and language are both Arabic, and so is the dating: the 7th of the month of Rajab 475 of the Hijrah era, corresponding to the 2nd of December, A.D. 1082.[123] Although these, as well as other details, especially the name of the lady buried at Leran, *binti Maymūn*, 'the daughter of Maymūn', suggest that this is a tomb of an Arabic lady and therefore is not strictly an example of Indonesian chronology, it has been noticed that the type of stone used for the tomb is one that is quarried locally, so that the inscription must have been written and dated in Java.

It is, however, an isolated example, because it is separated from the next by a gap of more than two centuries, when the Arabic inscription on the tomb of Maliku-'s-Salih, the first Sultan of Pasai, has a date corresponding to A.D. 1297.[124] In the fourteenth century there is the most interesting inscription from Trĕngganu near the east coast of the Malay Peninsula. Though outside Indonesia, it is also important with a view to developments in chronology in Indonesia. It is dated in the Hijrah era, but its precise date cannot be established with certainty owing to a lacuna immediately after the words *tujuh ratus dua*, indicat-

[122] Adolf Grohmann, 'Arabische Chronologie', *Handbuch der Orientalistik*, Erste Abteilung, Ergänzungsband II, Erster Halbband, Leiden/Köln, 1966, pp. 9-11 and *passim*.

[123] J.-P. Moquette, 'De oudste Mohammedaansche inscriptie op Java', *Verh. Eerste Congres voor de T. L. & V. van Java*, Weltevreden, 1921, pp. 391-399, and P. Ravaisse, 'L'inscription coufique de Leran à Java', *Tijdschr. Bat. Gen.*, LXV, 1925, pp. 668-703.

[124] J.-P. Moquette, 'De eerste vorsten van Samoedra-Pasè', *Rapp. Oudh. Dienst*, 1913, pp. 1-12.

ing the date.[125] If this is the end of the date the year is 702, but there are no fewer than twenty other possibilities, as the above mentioned words may be followed by either *puluh* or *-lapan*, alone or followed by one of the numbers from 1 to 9. The inscription gives, however, an additional detail, *viz.* the mention of the Year of the Crab, which limits the possibilities to only two, those corresponding to A.D. 1326 or 1386. The most interesting aspect of the date is the reference to the twelve-year cycle in which each year is named after a different animal. Twelve-year cycles are well known in Indian chronology where, however, the individual years are named after the twelve months, each preceded by *mahā-*, 'great' (*Mahācaitra* etc.). The animal names, however, are of Chinese origin and are first found in South East Asia in the tenth century in Cambodia.[126] The famous Buddha of 'Grahi' (Ch'aya in the Isthmus of Kra), dated at the end of the thirteenth century in the present writer's opinion,[127] is the earliest example in the Malay Peninsula and therefore likely to have been the direct source of the use of the twelve-year cycle in the Trĕngganu inscription. The dating thus shows a rather anomalous combination of different elements.

A different combination is found in another fourteenth-century inscription, the stone of Minye Tujuh (Aceh, Sumatra Utara). The year is A.H. 781 (A.D. 1380), the 14th of the month of Hajji (i.e. D̲ū 'l-ḥiǧǧa) but the week-day is given as *Śukrawāra* ('Friday') in its Sanskrit form (instead of Jum'at).[128] These examples represent a transitory phase in which the switch to the Muslim system of chronology was not yet complete. It is interesting that the only part of the date left over from the old system is the name of the week-day which, in ancient as well as in modern times, was more generally used than other elements of the date and therefore less easily replaced by a new name.

These examples show a kind of transitory phase between the old established and the more recently applied systems. As in other aspects of culture the transition to Islam in Java did not involve a complete abandonment of long established patterns. Thus, the old names of the week-days of Sanskrit origin, though obsolete, have not completely dis-

---

[125] For the inscription and the problems concerning its date cf. the publications mentioned in the present author's *Indonesian Palaeography*, 1975 in this same series, pp. 70 f.

[126] Cf. Éveline Porée-Maspéro, 'Le cycle des douze animaux dans la vie des Cambodgiens', *B.E.F.E.O.*, L, Fasc. 2, 1962, pp. 311-365.

[127] J. G. de Casparis, 'The Date of the Grahi Buddha', *Journ. Siam Soc.*, LV, 1967, pp. 31-40.

[128] W. F. Stutterheim, 'A Malay shā'ir in Old Sumatran characters of 1380 A.D.', *Acta Orient.*, XIV, 1936, pp. 268-279; C. Hooykaas, *Perintis Sastera*, transl. by Raihoel Amar gl. Datoek Besar, Groningen-Djakarta, 1951, pp. 73 f.; G. E. Marrison, 'A Malay Poem in Old Sumatran Characters', *Journ. Mal. Br. R.A.S.*, XXIV, Part I, 1951, pp. 162-165.

appeared. Outside central and eastern Java, and, of course, the island of Bali, however, the older system of chronology was never firmly established outside narrow court circles and so could disappear without leaving any visible trace.

CHAPTER TWO

INDONESIAN CHRONOLOGY FROM THE BEGINNING OF THE SIXTEENTH CENTURY

The conversion to Islam of large parts of the Indonesian archipelago before and during the sixteenth century led to widespread adoption of the Muslim era based on the Hijrah of the Prophet on the 15th of July, A.D. 622. The details of Muslim chronology have been fully discussed in another volume of this series so that no real discussion is required here.[1] Like the Śaka era and the chronology associated with it the Hijrah era also uses synodic months of the average length of 29,531 days but, unlike the Hindu year with its frequent intercalation of an extra month,[2] the Muslim year consists invariably of twelve synodic months of about 29.531 days and is therefore 354.367 days long or nearly eleven days shorter than the solar year.[3] No adjustments in order to equate the year with a whole number of months are made, so that each new year starts about eleven days later than the preceding one in relation to the solar year, moving, as it were, all through the (solar) year in cycles of about thirty-two years.[4] As the actual months consist, however, each of a whole number of days, adjustments have to be made. This is achieved by making some months last thirty days, others twenty-nine with, in addition, a complicated system of intercalary days.

The use of the Muslim calendar is essential for every Muslim because the recurrent Muslim duties and festivals, such as the fast of the month of Ramaḍān, are determined by this calendar. It was therefore generally adopted wherever populations or their ruling classes adopted Islam. In those parts of the Indonesian archipelago and the Malay Peninsula where the earlier system of Indian origin was unknown or was not firmly established there was no problem, but in Java, Bali and parts of Sumatra —in particular in central and eastern Java and Bali, which have yielded the vast majority of dated inscriptions and texts—the Śaka era had been in use since many centuries. The people, i.e. those classes that required a precise system of chronology, were accustomed to the Śaka era

---

[1] See note 122, p. 34.

[2] Approximately seven times in nineteen years or, to be more precise, 'about 369 mean added months in 1000 years' (W. E. van Wijk, *Decimal Tables*, 1938, p. 7).

[3] These figures are again those calculated by van Wijk; see the publication mentioned in the preceding note and the articles mentioned in note 12, p. 7.

[4] Adolf Grohmann, *Arabische Chronologie*, 1966, p. 13.

and were therefore unlikely to change their habits of dating. For some periods different eras may have been used concurrently, not only by different people in the same area, but sometimes even by the same people for different purposes, whereby the Muslim era would have been mainly limited to religious use, especially fo fix the dates of ceremonies, festivals etc., while the old era remained in use for most other purposes. (Such concurrent use of different systems of chronology can be compared with what we see in modern Indonesia, as well as in many other modern Muslim countries, where the use of the Muslim era is confined to religious matters, while the Christian era prevails for other purposes).

In Bali, where Islam never became prevalent, the Śaka era has remained in use till the present day, although it is gradually being replaced by the Christian era. On the other hand, in central and eastern Java, which from the middle of the sixteenth century came increasingly under the dominance of Mataram, the co-existence of the Śaka and Muslim eras resulted, probably in 1633,[5] to a curious, but very efficient, compromise whereby the counting of the Śaka years was continued but otherwise the length of the year (twelve synodic lunar months), the length and names of individual months, the names of weekdays etc. were changed to the Muslim pattern. In this manner there was no problem as it became possible to determine all the Muslim festivals without breaking the continuity with the past, at least in so far as the numbering of the years is concerned. Although the sources do not precisely define the measures that were taken when the Javanese era was established, it seems quite likely that the year A.H. 1043, which started on the 8th July, 1633 A.D., was simply given the number of the Śaka year 1555, current at that time—or to be more precise, which had elapsed: A.H. 1043 was equated with what had now become A.J. 1555. For all later dates one has to add the figure 512 to the year of the Hijrah era to obtain the equivalent year of the Javanese era.[6]

In addition, Javanese chronology from the middle of the seventeenth century shows two interesting special features. In the first place, the

---

[5] Some doubt has been expressed concerning the well known Javanese tradition ascribing the introduction of this Javanese era to Sultan Agung in 1633 (especially in the articles by Rouffaer and by D. G. Stibbe and C. Spat in the Encyclopaedia mentioned in note 3, p. 2) but such doubt seems unjustified. This matter, as well as some other problems concerning Javanese chronology are discussed by Dr. M. C. Ricklefs in Appendix 1 (pp. 223-236) to his *Modern Javanese Historical Tradition*, London, 1978.

[6] Tables showing the correspondences between the Muslim, Javanese and Christian dates are found in the Encyclopaedia articles mentioned in note 3, p. 2. A very convenient table illustrating the corresponding years in the three eras is found in Th. Pigeaud, *Javaans-Nederlands Handwoordenboek*, Groningen, no date, pp. IX-XI.

years go in eight-year cycles (*windu*) in each of which the individual years are named after a number of letters of the Arabic alphabet.[7] In Indian chronology cycles of twelve and of sixty years, both based on the movements of the planet Jupiter (Bṛhaspati), are well known, but there is no record of an eight-year cycle. Arabic chronology has a thirty-two years' cycle, the period during which the beginning of the Muslim year moves, as it were, through the solar year, as 33 lunar years approximately equal 32 solar years.[8] As the Javanese *windus* are again counted in groups of four, there is probably a connexion with the thirty-two year cycle of Arabic chronology.

In the second place, besides the Muslim year and its Javanese variant, there has been, in Java, at least since several centuries, a purely solar year, used mainly for agricultural purposes, called *Pranata Mangsa*. It is based on the division of the solar year into twelve months of unequal length, varying between about three and six weeks, ten of which are indicated by Javanese ordinal numbers, whereas the eleventh and twelfth months bear names derived from the names of the third and fourth months of the ancient Javanese and Indian year, viz. Ḍesta and Saḍa (from Jyeṣṭha and Āṣāḍha).[9] As this solar year has not been used for dating, but merely for regulating agricultural activities it would not have been mentioned even if it had been in use in ancient times. As a consequence its antiquity cannot be estimated. It seems, however, plausible that its introduction is directly due to the introduction of the Muslim chronology and its Javanese variant, neither of which bears any relation to the solar and seasonal phenomena on which the cycle of agricultural activities is based. The ancient system, based on luni-solar reckoning, did, although it contained a rather wide margin for error owing to the variable beginning of the ancient year. It is therefore possible that at some time after the introduction of the Muslim or Javanese year the ancient year was revived and adjusted for agricultural purposes. The fact that the months of this year are denoted as *mangsa* (from Sanskrit and Old Javanese *māsa*, 'month'), while the eleventh and twelfth months bear names derived from those of ancient months may give support to this view.

On the other hand, there is at least a possibility that the agricultural solar year can be traced back to ancient times. In an important article

---

[7] As spelt in Javanese the names are: 1. *alip*, 2. *ehe*, 3. *jimawal*, 4. *je*, 5. *dal*, 6. *be*, 7. *wawu*, 8. *jimakir* (Pigeaud, *op. cit.*, p. IX).

[8] For details about the different methods used in Arabic chronology cf. Adolf Grohmann, *l.c.*

[9] J. (L. A.) Brandes, 'De maandnaam Hapit', *Tijdschr. Bat. Gen.*, XLI, 1899, pp. 19-31, which mentions also earlier studies; cf. also the Encyclopaedia articles mentioned in note 3, p. 2.

J. L. A. Brandes discussed the name Hapit, which is often given to the two last months (Ḍesta and Saḍa) of the solar year, whereby the eleventh month is called Hapit Lĕmah and the twelfth Hapit Kayu.[10] He rightly argued that the term(s) can be found back in Old Javanese *Wariga* texts which give information about astronomical and astrological matters, partly for divination, partly also for the regulation of agricultural work. Unfortunately, it does not seem possible to date any of these texts within reasonably narrow limits, but Brandes has discovered a reference to the two Hapit months in a fairly accurately datable Old Javanese text, the *Arjunawiwāha*, composed in the reign of Airlangga, probably between c. 1028 and 1035.[11] The passage suggests that the two Hapit months correspond to the two months Ḍesta and Saḍa (Jyeṣṭha and Āṣāḍha) of the system outlined in the *Wariga* text, where they are the two last months of a year which starts in the month of Śrāvaṇa, corresponding to June/July of the Christian era (for instance, on the 6th of July in A.D. 1030).[12] Although this is approximately the time of the beginning of the *Pranata Mangsa* [13] in later times (on or shortly after the summer solstice) there is no real proof that the *Arjunawiwāha* passage alludes to a solar year; this seems nonetheless the most likely interpretation. The learned author of the text, Pu Kāṇwa, was apparently as familiar with the months used by the peasants as he was with the months of Vedic literature (Madhu and Mādhava) which are mentioned or alluded to in the same verse. It would therefore seem that the solar

---

[10] Brandes, *l.c.*, pointed out that *(h)apit* (also called *sĕla*, 'interim period') originally indicated the two months of relative slackness in agricultural activity between the rice harvest and the planting of other crops (*palawija*), a 'gap', as it were, in the agricultural year. The name *(h)apit*, literally 'stuck, jammed', may have been given on account of the position of the 'girdle' of Orion, which in these two months stays at or below the horizon at sunset, to re-emerge clearly after the summer solstice. Also some of the other months of the agricultural year are, according to the *Wariga* texts studied by Brandes, determined by the position of the Great Bear at sunset.

[11] The most recent discussion of the date of the *Arjunawiwāha* is that by Zoetmulder, *Kalangwan*, pp. 243-246 and the notes on p. 542. Zoetmulder is unconvinced by the arguments put forward by C. C. Berg to cast doubt on the authenticity of the stanza containing the date of the *kakawin*.

[12] Brandes' translation of the words *Madhumāsakāla tĕka ring Hapit* as 'the Honey months and (the months of) Hapit' is not precise as the passage suggests the month of Madhu (i.e. Caitra) up to (*tĕka*) Hapit (i.e. Jyaiṣtha and Āṣāḍha), in other words the first four months of the lunisolar year, corresponding, in northern India, to spring (*vasanta*) and summer (*grīṣma*). These are the driest months of the year just preceding the rainy season, and the longing of the plants for rain is an apt simile for the rise of passion. Brandes' note commenting on the 'foolish and slavish' imitation of an Indian prototype is not quite justified for, although the rainy season does not set in until several months later in East Java, this is also a dry period and one in which there is little agricultural activity.

[13] There were, however, different systems to determine the precise beginning of the year and the lengths of the months. As indicated in the supplement of 1927 to the *Encyclopaedia* (p. 405), the beginning of the year was, in 1855, fixed on the 21st of June, and the length of each individual month was fixed at definite numbers of days, all between three and six weeks. This point will be fully elaborated by Dr. Ricklefs in his forthcoming publication mentioned in note 5 above.

year is, in fact, old, but its use was limited to peasant classes; it therefore did not appear in the official documents or in literary works, except in this particular case. It is also easy to understand why this agricultural year became more important in the Mataram period. In contrast to the Śaka era, in which the times of the main agricultural activities could be easily fixed (it could never be out of tune with the solar year by more than a month), the Muslim year and its Javanese variant were irrelevant for agriculture.

As to the origin of the Pranata Mangsa it may be argued that solar years are well known in India, too, so that it is not impossible that this particular solar reckoning is based on Indian chronology. It is true that solar years in ancient India normally started near the vernal equinox, as was the case with lunisolar reckoning, but not, of course, at the first New Moon after True Mīna Saṃkrānti but at the Sun's entrance into Meṣa (Aries).[14] There is, however, also evidence for solar years starting with the Sun entering into Leo (*siṃhādi*) and Virgo (*kanyādi*), in particular the Kollam era (*kollam āṇḍu*) of southern Kerala.[15] The former (*siṃhādi*) begins with the month of Śrāvaṇa, exactly like the Javanese solar year.[16] In India this is traditionally the beginning of the rainy season (*varṣa*) and this holds true for Kerala where the first rains normally fall in June. This is, of course, the natural beginning of the cycle of agricultural activities, and the adoption of this form of solar reckoning makes good sense in Kerala. In central or eastern Java, where the west monsoon does not set in until at least three months later, the beginning of the year in June/July is more difficult to understand, unless it was introduced from India. The latter is therefore a distinct possibility, although the balance of arguments—in particular its association with agriculture, where Indian influence seems almost completely absent, as well as the Indonesian name Hapit given to two of its months and Indonesian numerals to the others—is clearly in favour of the view, expressed by Rouffaer and others, that the Pranata Mangsa is originally Indonesian.

Apart from the Javanese era Indonesian chronology from the beginning of the sixteenth century shows few interesting aspects that require a detailed discussion. The Muslim era based on the Hijrah on July 16th,

---

[14] Cf. W. E. van Wijk, *Decimal Tables*, the graph facing page 3, under 21[1], 21[2], and 22.

[15] Van Wijk, *op. cit.*, pp. 4 f.

[16] D. C. Sircar, *Indian Epigraphy*, pp. 269 f.—Whereas in northern Kerala the months in this era are named after the *rāśis*, we see that in southern Kerala the names of the lunar months are used. It is striking that this method is also followed in the Javanese Pranata Mangsa, at least for the months which are not counted by ordinal numbers (i.e. the first, eleventh, and twelfth).

A.D. 622, was in wide-spread use all over the Indonesian archipelago and the Malay peninsula, but mainly in association with religion. It is therefore used on Muslim tomb stones and in Muslim religious literature. In Malacca, Aceh, Bantĕn and other states where Islam was the basis of political life the Muslim era was in general use as the official system of chronology. As it shows no special Indonesian features there is no need for any discussion here.

The Christian era, first introduced by the Portuguese in the beginning of the sixteenth century, long remained a foreign era as its use was confined to the Western settlers and their relatively small numbers of Indonesian and Chinese converts to Christianity. During the nineteenth and the first half of the twentieth century it spread to most parts of the Archipelago. It was during that time the official chronology of the Netherlands East Indies government and its agencies. It also became the official chronology of the Republic of Indonesia, founded on the 17th of August, 1945. The Christian era shows no more special Indonesian features than the Muslim era and therefore requires no discussion.

As mentioned earlier, the Śaka era has remained in somewhat limited use in Bali where it is still widely known. We saw earlier that the inscriptions generally give few astronomical details in comparison with Old Javanese datings but have, in addition to the six-, five-, and seven-day weeks, also a three-day week. In more recent times there are, in addition, four-, eight-, and nine-day weeks. All these different weeks are not used in normal dating but are important for divination. One 'calendar' indicating all these week days, in fact a wooden calender plank (*tika*), has been studied by W. O. J. Nieuwenkamp.[17] It consists of thirty vertical columns, divided into 210 little squares by seven horizontal lines. The squares are filled with different types of symbols indicating not only different weekdays but also the kind of activities that should or, on the contrary, should not be undertaken on such days.

Apart from such regular systems of chronology there have been till recent times—and have not perhaps completely disappeared at the present time—some less elaborate, often rather incomplete, types of chronology in different parts of the Indonesian archipelago. Most of these systems are described by Rouffaer in a long article on Chronology in the *Encyclopaedie van Nederlandsch Oost-Indië* of 1905 and by D. G. Stibbe, C. Spat and E. M. Uhlenbeck in the supplement to the same Encyclopaedia of 1927.[18] Some types of chronology have been studied

[17] W. O. J. Nieuwenkamp, 'Een Balineesche Kalender', *Bijdr. Kon. Inst.*, 69, 1914, pp. 112-126.
[18] See note 3, p. 2.

to a greater depth, e.g. that of the Landak-Dayak of West Kalimantan, described by M. C. Schadee.[19] Although this chronology has been described as 'primitive' it is partly based on keen observation of the sky in different parts of the year. Thus, it includes some very ingenious methods to estimate the height of certain stars above the horizon in order to determine the right time for various agricultural activities. The study of these methods as well as similar methods applied in other parts of the archipelago falls, however, outside the scope of this study.

On the whole it may seem that, in comparison with earlier times, there has been a remarkable decline in the value and the accuracy of chronology in Indonesia after about A.D. 1500. This is especially noticeable in historiography. Old Javanese works, such as the *Pararaton* and the *Nagarakrĕtāgama*,[20] contain numerous dates, in figures, in chronograms or in both. The former has 52 dates in all, the latter 40, so that all important events are firmly fixed in time. The later historical works, composed after c. A.D. 1500, contain generally few dates or even none at all. Thus, such a basic text as the *Sĕjarah Mĕlayu*, 'Malay Annals' as the title is usually, though incorrectly,[21] rendered in English, gives no dates at all, not even for the foundation of Malacca. It is true that this text is by no means devoid of chronology in so far as the length of the reigns of the sultans is regularly recorded in years, but it does not state the regnal years of the different rulers in which important events took place. Despite the detailed account of the Portugese attack on Malacca no date is given, except that it is clear that it took place during the reign of Maḥmūd Shāh. Most of the indications of time are vague ('it happened once that . . .', 'after some time', 'one day', 'after a while', etc.) and this, no less than the actual description of different events, lends the work the character of a story or rather a collection of stories. The same applies to numerous other works of historical value, such as the *Hikayat Banjar*,[22] the *Hikayat Patani*,[23] the *Hikayat Malèm Dagang*[24]

[19] M. C. Schadee, 'De Tijdrekening by de Landak-Dajaks in de Westerafdeling van Borneo', *Bijdr. Kon. Inst.*, 69, 1914, pp. 130-139.

[20] Th. Pigeaud, *Java in the Fourteenth Century*, Vols. I-V, The Hague, 1916; J. L. A. Brandes, *Pararaton*, 2nd ed., prepared by N. J. Krom, with contributions by J. C. G. Jonker, H. Kraemer and R. Ng. Poerbatjaraka, The Hague-Batavia, 1920.

[21] R. O. Winstedt, 'The Malay Annals or Sĕjarah Mĕlayu', *Journ. Malay Branch R.A.S.*, XVI, 1938, pp. 1-226; C. C. Brown, 'Sejarah Melayu or Malay Annals', translation, *ibid.*, XXV, 1952, parts 2 & 3, pp. 7-276.

[22] J. J. Ras, *Hikajat Bandjar. A Study in Malay Historiography*, The Hague, 1968. (*Bibliotheca Indonesica*, 1).

[23] A. Teeuw & D. K. Wyatt, *Hikayat Patani. The Story of Patani*, The Hague, 1970 (*Bibliotheca Indonesica*, 5).

[24] H. K. J. Cowan, *De "Hikajat Malem Dagang"*, The Hague, 1937.

and many others. The *Babad Tanah Jawi*, on the other hand, contains some, though relatively few, dates,[25] but some other texts, such as the *Sĕrat Kaṇḍa* and, in particular, the *Babad ing Sĕngkala*, give many dates.[26] It would therefore seem that at least some of the later Javanese historical texts contain many accurately dated events. If one considers that there are also many Old Javanese texts which are completely devoid of dates —in fact, all the *kidungs* belong to this group—then it follows that there is, in this respect, no striking difference between ancient and modern Java. In both periods there were probably different traditions existing side by side: a real historical tradition with emphasis on an accurate record of (some aspect of) the past and a more literary tradition in which 'stories' of different kinds, often illustrating particular moral or political values, carried greater weight than the sober historical narrative. There were, of course, different shades of emphasis between these two genres. Accurately dated documents, very similar to some of the Old Javanese, Old Sumatranese and Old Balinese inscriptions, are found until relatively recent times. Among the most interesting examples are the *piagĕms*, all in modern (or rather pre-modern) Javanese, issued by the sultans (or high dignitaries acting in the sultan's name) of Palembang, South Sumatra.[27] They are all accurately dated, mentioning the 'Śaka' year (in fact, the year of the Javanese era, i.e. the Śaka era with, from A.D. 1633, the length of the Muslim year), the year of the eight-year *windu*-cycle, the week-day of the seven-day week, ordinal number of the date and the name of the month. The name of the week-day can be used to check the correctness of the reading of the date as a whole. Similarly, one would have thought that the mention of the *windu* cycle could be used to check the correctness of the reading of the figures of the year, but here there is a difficulty as the cycle does not seem to have been used in the same way as in Java. Only in one case, a *piagĕm* of A.J. 1656 (A.D. 1731), year *Be*, the name of the year agrees with that given to this year in Java, but in all the other cases there is disagreement. As it seems unlikely that so many dates should be incorrect it would follow that the Palembang scribes used a different system of naming the years of the *windu* cycle.[28] Owing to the relatively small number of inscriptions (fewer

---

[25] W. L. Olthof, *Babad Tanah Djawi*, 2 Vols., The Hague, 1941.

[26] Cf. M. C. Ricklefs, *Jogjakarta under Sultan Mangkubumi*, 1749-1792, p. xxiii. J. L. A. Brandes, *Pararaton*, 2nd ed. p. 214.

[27] J. L. A. Brandes, 'Nog eenige Javaansche piagĕm's uit het Mohammedaansche tijdvak, afkomstig van Mataram, Banten en Palembang. III. Palembang', *Tijdschr. Bat. Gen.*, XXXIV, 1891, pp. 605-623; XXXVII, 1894, pp. 121-126; XLII, 1900, pp. 131-134 and 491-507.

[28] Brandes, *Tijdschr. Bat. Gen.*, XLII, 1900, p. 493.

than twenty in all) it has not yet been possible to define this different system.

Another part of Indonesia where historiography flourished from the seventeenth century is South Celebes. Most of the historical texts, however, show a similar attitude towards precise dates as Malay historiography. Thus, one of the foremost authorities in this field, J. Noorduyn, points out that, although the manner in which developments and events are presented inspires confidence in their accuracy, dates are almost always missing and events are connected by such phrases as 'after some time' or 'after some years'.[29] Noorduyn further suggested that the absence of dates in the more or less official chronicles may be due partly to literary considerations and partly to the 'possibility that the precise dates were thought to be sufficiently and more properly preserved in other kinds of writing, as in the so-called diaries.' These Macassarese and Buginese diaries are among the most valuable sources for the history of Celebes (and occasionally other areas) in the seventeenth and eighteenth centuries. They are unique in several respects, not in the least on account of the precise dates which precede all the information that is recorded. In many cases the dates are given in both the Muslim and the Christian era, sometimes, however, with errors, which can be attributed to later scribes.

In the light of the examples from Java, Palembang and South Celebes it seems wrong to conclude to a general decline in the importance of chronology after the sixteenth century. It would rather seem that, as in ancient Indonesia, strict adherence to accurate chronology in documents or in historical narrative depended on the nature of these texts rather than on the period in which they were written. Documents of legal value, whether in ninth-century Java or in eighteenth-century Palembang, had to be precisely fixed in time. The same applied to diaries and a certain type of chronicles, such as the *Babad ing Sěngkala* and, for instance, some of the recently published Malay texts concerning the history of Palembang mainly in the eighteenth and nineteenth centuries.[30] Other types of chronicles, however, were not primarily concerned with precisely recording events in their chronological sequence, but laid

---

[29] J. Noorduyn, 'Some Aspects of Macassar-Buginese Historiography' in: *Historians of South East Asia*, edited by D. G. E. Hall, London, 1961, pp. 29-36; 'Origins of South Celebes Historical Writing' in: *An Introduction to Indonesian Historiography*, edited by Soedjatmoko and others, Ithaca, 1965, pp. 137-155; 'Tentang Asal-mulanja Penulisan-Sedjarah di Sulawesi Selatan', *Madj. Ilmu-Ilmu Sastra Indonesia*, III, 1966, pp. 212-232 (the original Indonesian version of the public lecture given in 1964).

[30] M. O. Woelders, *Het Sultanaat Palembang*, 1811-1825, The Hague, 1975 (= *Verh. Kon. Inst.*, 72), pp. 67-413.

particular emphasis on the origin of states or royal families or concentrated rather on the literary presentation of stories connected with such states or royal families. It has rightly been emphasized that some of the texts of this kind were composed to increase, by literary magic, the splendour of the dynasty. These different types of chronicles are not always clearly defined, and there are numerous possibilities of combination of different features, as well as of various shades of emphasis. Even such texts as the *Babad Tanah Jawi*, which may present some of the clearest examples of literary magic, are not totally devoid of dates. Generalizations about the decline of interest in chronology, in particular in precise dating, in Indonesia after the beginning of the sixteenth century are not supported by the available evidence, which suggests that the importance attached to chronology depended above all on the type of texts and, to some extent, on the particular part of Indonesia where the text was composed.

APPENDIX I

YEARS, MONTHS ETC.

A. YEARS

(a) Śaka Era:

1. *Beginning of the year.* The beginning of the year has to be calculated for each individual year. According to the most common Indian system, which was invariably applied in ancient Indonesia, the new year was to begin at the first new moon after the sun entered the zodiac sign of Pisces (True Mīnasaṃkrānti, often abbreviated T.M.S.). It is not necessary to prepare tables showing the beginning of each individual year from the seventh century. Such tables can be found in R. Sewell and S. B. Dikshit, *The Indian Calendar*, London, 1896. In addition, the initial dates of each year can easily be calculated by means of the tables prepared by W. E. van Wijk on the basis of the *Sūryasiddhānta*, especially those in *Acta Orient.*, II. 1924, pp. 235-249 (Table VI: Julian Date of True Mīna Saṃkrānti, and VIII: ☽ — ⊙, *i.e.* number of days from T.M.S. to the first new moon) or the somewhat simplified tables of his publication of 1938. The sum of the final figures shows the number of days after the beginning of the year in our era (*i.e.* Śaka Era + 78). It should be added that all dates have first to be converted into the Kaliyuga Era (by adding 3179 to the Śaka year or 3101 to the A.D. year) before van Wijk's tables can be used. By means of a pocket calculator the entire operation takes no more than one or two minutes.

2. *Length of the year.* As seen above on p. 8, the figure obtained from Table VIII in van Wijk's article shows whether the year is one of twelve or thirteen months and, in the latter case, the name of the month that is to be repeated (see also the table on p. 240 of the same article).

(b) Muslim Era:

As Muslim chronology has been discussed by A. Grohmann in another volume of this *Handbuch* (*Arabische Chronologie*, Leiden, 1966) no details are required here. Tables for the conversion of any year of the Hijrah Era into the corresponding A.D. years are easily available (e.g. in Th. Pigeaud, *Javaans-Nederlands Handwoordenboek*, Groningen, no date, pp. IX-XI).

(c) JAVANESE ERA:

For all details regarding the complexities of this era the reader is referred to Appendix I and II (pp. 223-244) of the new book by M. C. Ricklefs, *Modern Javanese Historical Tradition*, London, 1978. A convenient table for the conversion of dates of the Javanese era is given by Pigeaud, *loc. cit.* Owing to an erroneous assumption of Rouffaer (see Ricklefs, pp. 293 f.) Pigeaud's list starts actually eight years before 1633, the year of the foundation of the era.

(d) JAVANESE SOLAR YEAR:

More properly called 'quasi-solar' or 'agricultural year' (Ricklefs, pp. 229-231). For all details the reader is again referred to Dr. Ricklefs' publication. The same scholar discusses the Javanese eight-year *windu* cycle on p. 234.

B. MONTHS

(a) ŚAKA ERA:

The names of the Indian months are well known and can be found enumerated (with some common variants) in van Wijk, *Acta Orient.*, II, 1924, Table VII (p. 248); Renou-Filliozat, *L'Inde Classique*, II, 1953, pp. 732 f.; A. L. Basham, *The Wonder that was India*, reprint 1961, pp. 492 f. In Indonesia, however, there are some minor differences in the form and spelling of the names. The list which follows is based on that given by Damais, *B.E.F.E.O.*, XLV, Fasc. 1, 1951, p. 11:

1. Caitra,
2. Waiśākha,
3. Jyeṣṭha,
4. Āṣāḍha,
5. Śrāwaṇa,
6. Bhadrawāda,
7. Asuji,
8. Kārttika,
9. Mārgaśira,
10. Poṣya,
11. Māgha,
12. Phālguṇa.

Once the corresponding A.D. date of the beginning of the year has been established the beginning of each following month can easily be calculated by the addition of the required multiple of 29.531. If, however, an extra month has to be intercalated between the beginning of the year and the month for which the calculation is to be made, it is, of course, necessary to add an extra figure of 29.531. To avoid the possibility of a mistake the use of van Wijk's Table XIII, Parts I and II (*Acta Orient.*, IV, 1925, pp. 79 f.) is recommended.

(b) MUSLIM AND JAVANESE ERAS:

For a list of the Arabic months with their variants see especially A. Grohmann, *Arabische Chronologie*, Leiden/Köln, 1966, pp. 12 ff. In Indonesia, however, many different forms of the Arabic names are used, while in some cases, especially in Java, quite different names are used. Thus, instead of the Arabic name for the third month, *Rabī'u 'l-'awwalu*, the name *Mulud*, 'Birth (of the Prophet)', is substituted, while the name of the month of Ramaḍān is often replaced by Puasa in Indonesian or Påså in Javanese, both words indicating the fast during this month. The names and lengths of the different months as used in modern and pre-modern Java are listed by Ricklefs, *op. cit.*, p. 229.

(c) JAVANESE 'SOLAR' YEAR:

For a list of the names and lengths of the months of this mainly agricultural year see M. C. Ricklefs, *op. cit.* pp. 229-231. Different systems have been in use, for which cf. the articles 'Tijdrekening' in the first and second editions of the *Encyclopaedie van Nederlandsch-Indië* of 1905 and 1927 (the original article was by G. P. Rouffaer) and J. L. A. Brandes, 'De Maandnaam Hapit', *Tijdschr. Bat. Gen.*, XLI, 1899, pp. 323-343. What all these systems have in common is the use of twelve months of very unequal lengths (between twenty-three and forty-three days) and a starting point in early summer. The system was regularized in A.D. 1855 with the result listed in the second edition of the *Encyclopaedie* on p. 405 and, more clearly, in the work by Ricklefs on p. 230.

C. WEEKS

The Old Javanese inscriptions invariably state the names of the days of each of the three weeks, *i.e.* of the six-day, five-day and seven-day weeks (always in that order), often (especially from the second quarter of the tenth century) supplemented by the name of the *wuku*. These cyclic elements (the length of the cycle is 210 days) have especially been studied by Damais and exhaustive details can be found in the Appendices 1-9 of his 'Études d'Épigraphie Indonésienne', IV in *B.E.F.E.O.*, XLVII, Fasc. 1, 1955, pp. 252-283. The tables also include the Balinese three-day week. In Appendix 2 Damais gives a list of the initial dates of the 210 day cycle in all the years from the beginning of the Christian era to A.D. 2240! On the other hand, the table of Appendix I, though admirably accurate, is not easy to use. Some-one looking for, for example, the combination *ha*, *wa*, *bu* (Haryang, Wagai, Wednesday) has to go through most of the table before finding the combination under No. 158. I therefore added my own table (Appendix III) by which any combination can be found in a few seconds. With the help of this table one can immediately find the *wuku* name, if only the three week-days are given, or the missing week-day, if two out of three and the *wuku* are mentioned. The main purpose of the table, however, is to enable the reader to convert any combination of week-days into a figure from 1 to 210, showing the number of days from the beginning of a cycle (listed in Damais' Appendix 2) to the date on which the combination corresponding to the number occurs. I also used the table as a means of checking the distance in time between a 'given' date and one the reading of which is not beyond doubt. If, for example, the former gives Haryang, Wagai, Wednesday, *i.e.* No. 158 and the latter has the week days Paniron, Wagai, Thursday, *i.e.* No. 208, then it follows that the latter, of not quite certain date, must fall 50 days after the former date which is certain, or 50 + 210n days.

D. OTHER PARTS OF THE DATES

As to the other constituent parts of the dates in inscriptions it may be useful to give a few more precise data.

(1) Each month is divided into 30 TITHIS, 'lunar days', exactly one thirtieth of a synodic month or .984 of a mean day. Once in about two months it therefore happens that a *tithi* starting just after sunrise ends before sunset. Such a *tithi* is 'lost' (*kṣaya-tithi*). As a consequence one day of the month, in which the lost *tithi* occurs, is also 'lost', so that the month is, in fact, one of 29 days.

The *tithi*s are divided into two groups of fifteen, the first belonging to the waxing moon (*śukla-pakṣa*), the second fifteen to the waning moon (*kṛṣṇa-pakṣa*). They are given Sanskrit ordinal numbers in the feminine, as they qualify the word *tithi*. The only exception to this rule is the first *tithi* of each *pakṣa*, which is called *pratipāda* (Sanskrit *pratipadā*). The Sanskrit names *pūrṇimā* (full moon) and *amāvāsyā* (new moon) are not used in Indonesia, and are replaced by *pañcadaśī śuklapakṣa*, resp. *kṛṣṇapakṣa*. So the list of *tithi*s for each of the halves of the month is as follows:

1. *pratipāda*
2. *dwitīyā*
3. *trětīyā*
4. *caturthī*
5. *pañcamī*
6. *ṣaṣṭī*
7. *saptamī*
8. *aṣṭamī*.
9. *nawamī*
10. *daśamī*
11. *ekādaśī*
12. *dwādaśī*
13. *trayodaśī*
14. *caturdaśī*
15. *pañcadaśī*.

(2) A KARAṆA is half a *tithi* and therefore a mean *karaṇa* equals .492 days. The names of the *karaṇas* have already been given on p. 23 and do not need to be repeated here. It should, however, be noted that, because the *karaṇas* go in cycles of seven, the names occurring as those of the first *karaṇas* of the day in the first half of the (seven-day) week become the second *karaṇas* in the second half of the week. The inscriptions only seem to record the first *karaṇa*, *i.e.* that which is current at sunrise of the month mentioned. The first and the three last *karaṇas* of each day have special names (see p. 23). These are the four *karaṇas* of the new moon day, including the immediately preceding and following periods. A useful table showing the numbers of *karaṇas* and *tithis*, their names, the precise times of their beginning, etc., as well as the manner in which *tithis* and *karaṇas* correspond, is given by van Wijk in *Acta Orient.*, IV, 1925, pp. 73 f.

(3) NAKṢATRA, DEWATĀ and YOGA. Lists of the *nakṣatras*, which in India have been an essential part of the date since Rigvedic times but are given regularly in Indonesia only from the second quarter of the tenth century, are easily available. For the names in Vedic literature see P. V. Kane, *History of Dharmaśāstra*, Vol. V, Pt. I, Poona 1958, pp. 498-508, for those in classical times see *e.g.* W. E. van Wijk, *Acta Orient.*, IV, 1925, Second Part of Table X on p. 77. Van Wijk records three different systems: (a) 'equal space', whereby the ecliptica is divided into 27 equal parts, (b) that of Gārga with unequal space between the different asterisms, more or less in accordance with the space occupied by each asterism, and (c) that of Brahmagupta, also with unequal spacing but with rounded-off figures to the nearest whole or half number of days. Unlike the two other systems that of Brahmagupta counts with 28 *nakṣatras*, with the additional one, Abhijit, at the 22nd place. This additional *nakṣatra*, mentioned already in the *Atharvaveda* but absent from the *Jyotiṣavedāṅga* and most later texts, was apparently added by Brahmagupta to account for the fraction .322 of the complete period of a sidereal month, *viz.* 27.322 days (the 27 other *nakṣatras* totalling precisely 27 days). No example of Abhijit is known to me from inscriptions in Indonesia which, owing to the limited nature of the evidence, is no proof but may establish some degree of probability that Brahmagupta's system was not used in Indonesia. It is, however, impossible to choose between the other two.

For the sake of convenience I am giving a list of the *nakṣatras*, which is, apart from a few minor differences in spelling, that supplied by van

Wijk; the list of the corresponding *dewatās*, based on the *Jyotiṣavedāṅga*, is taken from P. V. Kane, *op. cit.*, pp. 501-504. These names are useful as they enable us to check the correctness of the reading of the name of the *nakṣatra*, which itself provides a valuable check as to the correctness of the reading of the date as a whole.

1. Aświnī	Aświnau	15. Swāti	Wāyu
2. Bharaṇī	Yama	16. Wiśākhā	Śakra
3. Kṛttikā	Agni	17. Anurādhā	Mitra
4. Rohiṇī	Prajāpati	18. Jyeṣṭ(h)ā	Indra
5. Mṛgaśiras	Soma	19. Mūla	?
6. Ārdrā	Rudra	20. Pūrwāṣāḍha	Āpaḥ
7. Punarwasu	Aditi	21. Uttarāṣāḍha	Wiśwedewāḥ
8. Puṣya	Bṛhaspati	22. Śrawaṇā	Wiṣṇu
9. Āśleṣā	Sarpā(ḥ)	23. Dhaniṣṭhā	Wasawaḥ
10. Maghā	Pitaro(-aḥ)	24. Śatabhiṣaj	?
11. Pūrwa-phalgunī	Bhaga	25. Pūrwabhadra-wāda	Ajapāda
12. Uttara-phalgunī	Aryaman	26. Uttarabhadra-wāda	Ahirbudhnya
13. Hasta	Sawitṛ	27. Rewatī	Pūṣa(n)
14. Citrā	Twaṣṭṛ		

The Old Javanese inscriptions give examples of most, though not all, of the *nakṣatras* and *dewatās*, and show a few differences from the list given here. Some of these are merely differences in spelling (as in *Aświno* for *Aświnau*, as the *au* of Sanskrit words is almost always replaced by *o* in Old Javanese) but there are also more substantial differences. Thus, *Bhaga*, the deity protecting Pūrwa-phalgunī, originally one of the forms of the Sun god (also Aditi, Sawitṛ, Mitra and Pūṣan clearly have solar connotations), must later have been interpreted as meaning 'vulva', for in at least one Old Javanese inscription the name of the protective deity is given as Yoni-dewatā. The conception of Yoni as a deity has clear Tantric connotations. In other cases synonyms are used. Thus, the deity protecting Swāti is given as Pawana instead of Wāyu, the protector of Śrawaṇā is Hari instead of Wiṣṇu, that of Kṛttikā is Dahana instead of Agni, and that of Jyeṣṭ(h)ā is not Indra but Śakrāgni. This may appear as merely a play of words, but the last example is interesting because Śakra and Indra, though often regarded as names of the same god, are protectors of different *nakṣatras* in the above list. One case that is difficult to understand is the replacement of Bṛhaspati, as lord of Puṣya, by just Dewa. One possibility is that Dewa here stands for Dewaguru, a well known name of Bṛhaspati. There are a few minor differences (thus, the Old Javanese inscriptions give Pitṛ in the singular as protector of Maghā instead of the usual plural form Pitaraḥ in the Sanskrit texts). Such details may not seem very interesting, but they may be helpful in any study of the Indian sources used by the Indonesians.

Finally, the *yoga* is given in almost all inscriptions from the tenth century on. Lists are easily available as the *yoga*, like the *nakṣatra*, is the normal constituent of a *pañcāṅga*. The names are given, for instance, by van Wijk in the Second Part of Table XI in *Acta Orient.*, IV, 1925, p. 78. I have found no differences whatever in the Old Javanese inscriptions. The *yoga* is also easy to check by means of the First and Second Parts of Van Wijk's Table XI, the second column on pp. 75 f. and 78.

(4) The RĀŚI, 'sign of the zodiac', is, as we already saw on p. 22, not regularly mentioned in Old Javanese inscriptions before about the middle of the twelfth century. This may seem surprising because (a) the zodiac signs must have been well known as their knowledge is essential for determining whether there is an intercalary month or even for fixing the beginning of the year, and (b) the *rāśi* always was one of the essential criteria in Indian astrology to determine whether a date was auspicious or otherwise. As to (a) we have to conclude that the Indonesian astronomers, though using the *rāśi*s for their calculations, saw no reason to add this detail to their dates, presumably because (b) it was not used to determine whether a date was auspicious. The emphasis in astrology is always laid on combinations of week-days, *wuku*s and other cyclic elements (such as the *windu* cycle). Yet, from the second half of the twelfth century the *rāśi* is regularly mentioned in inscriptions, which may suggest that new influences from India, apparent also in other fields (such as the emergence of Vaiṣṇavism as an important religious movement), made themselves felt. There may be some connexion with the so-called *prasen*s ('zodiac beakers'; see pp. 32 f.), which are, however, not attested until two centuries later. Lists of the twelve *rāśi*s are easily available. See especially the most detailed and scholarly discussion by P. V. Kane, *op. cit.*, pp. 561-580 and *passim*. Tables to calculate the precise time of the beginning of each *rāśi* (*i.e.* the *saṃkrānti*: the time when the sun enters the constellation) are given by van Wijk in *Acta Orient.*, II, 1922, p. 248. Unfortunately, the *rāśi*s mentioned in Old Javanese inscriptions raise some difficulties. Although in some cases we find the *rāśi* agreeing with the date (e.g. *wṛścika*, 'the Scorpion', associated with the month of Kārttika in *O.J.O.*, LXVI and LXXIX; *dhanu*, i.e. *dhanus*, 'the Bow, Sagittarius', associated with Mārgaśīrṣa in *O.J.O.*

LXXI), there are many cases where there is disagreement. In some of these it is possible that the *lagna*, 'horoscope', 'the constellation in which the rising planet (Jupiter?) appears', is meant. Thus we find *wṛścika* combined with Āṣāḍha in *O.J.O.* LXXIV, *mīna* with Bhadrawāda (Bhādrapada) in *O.J.O.* LXXII, and *mithuna* with Mārggaśira (Mārggaśirṣa) in *O.J.O.* LXXXIII. Further research is needed.

For the sake of convenience I am here giving the list of *rāśis* as found in the above mentioned works by Kane and van Wijk:

1. Mīna (Pisces),
2. Meṣa (Aries),
3. Vṛṣabha (Taurus),
4. Mithuna (Gemini),
5. Karka (Cancer),
6. Siṃha (Leo)
7. Kanyā (Virgo),
8. Tulā (Libra),
9. Vṛścika (Scorpio),
10. Dhanu(s) (Sagittarius),
11. Makara (Capricornus),
12. Kumbha (Aquarius).

(5) Finally, the MUHŪRT(T)A, 'auspicious moment (to carry out a ceremony, to set off on a journey, etc.); also 1/30th of a day, *i.e.* 48 minutes', is mentioned regularly only from the middle of the thirteenth century. As one may have expected, not all the *muhūrta* names are actually found in inscriptions. All the foundations which are mentioned in inscriptions seem to have taken place in daylight, for none of the fifteen 'hours' of the night has hitherto been found in inscriptions. Of the fifteen day-time hours only five are attested, viz. Raudra, Śweta, Wairāja, Wijaya and Saumya, the first three in the morning, the two others in the afternoon. Saumya (from 16.24 to 17.12 hours, if the mean system is followed) occurs at least three times. It was, quite understandably, a popular hour to conduct the ceremonies in connexion with the transfer of land, usually for religious purposes. It is a pleasant time of the day in East Java. The formal arrangements would have been completed before dusk, which was the time to start the festivities, which included not only eating and drinking (some inscriptions even give lists of dishes and of drinks), but also music, dance and, at least in some cases, if not in all, shadow-play. The festivities no doubt continued till dawn.

For a complete list of the *muhūrtas* see P. V. Kane, *op. cit.*, note 788 to p. 539.

APPENDIX II

GRADUAL LENGTHENING OF THE EXPRESSION OF DATES IN OLD-JAVANESE INSCRIPTIONS

The adjoining chart is meant to illustrate the gradual lengthening of the dates in Old Javanese inscriptions by the addition of new details. The chart is somewhat over-simplified in that it is confined to what appears to have been the normal formulation of the date in each period. Within each period, however, one finds a certain amount of flexibility. On the one hand, one finds a substantial minority of cases in which one or even more elements of the date, expected for the period, are omitted; on the other hand, there are a small number of examples where one or, more rarely, several elements of the date, usual in later periods, are, as it were, anticipated. Such examples illustrate the same kind of flexibility which is noticeable in the formulation of royal titles, privileges granted to donees, kinds of people excluded from consecrated land etc. In all these cases there is a clear tendency of gradual increase of the numbers of details mentioned in inscriptions: the later the inscription, the greater the number of details. But in respect of such details, too, there are occasional examples of very lengthy enumerations in earlier inscriptions and relatively short formulations in the later ones. The general tendency is, however, clear. The following examples show what may be regarded as the standard manner of dating:

I. Before A.D. 900: the Ngabean Copper-Plate Inscription of Kayuwangi, dated A.D. 879, *O.J.O.* XII, line a 1.
II. From c. 900 to 1000: the Randusari Copper-Plate Inscription of Balitung, dated A.D. 905, *Inscr. Ned.-Indië*, I, 1940, p. 4, line 1 b 1.
III. From c. 1000 to 1250: the Cĕkĕr (Tjeker) Stone Inscription of Kāmeśwara, dated A.D. 1185, *O.J.O.*, LXXII, Front, lines 1-3.
IV. After 1250: the Pĕnanggungan Copper-Plate Inscription of Kĕrtarājasa, dated A.D. 1296, *Inscr. Ned.-Indië*, I, 1940, p. 38, lines 1 b 1-3.

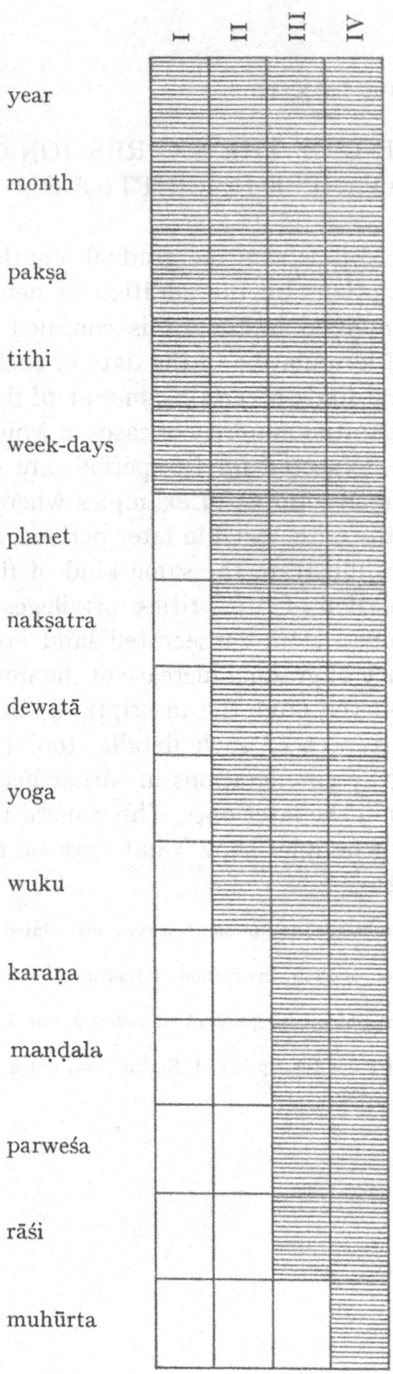

I ......Before c. 900.

II ......c. 900-1000.

III ......c. 1000-1250.

IV ......After c. 1250.

APPENDIX III

TABLE TO CONVERT ANY COMBINATION OF WEEK-DAYS AND WUKU INTO A FIGURE FROM 1 TO 210

The use of this Table has already been explained in Appendix I. The *wukus* are represented by Roman numerals according to the following table of correspondence:

I	Sinta	XVI . . .	Pahang
II	Landĕp	XVII . . .	Kuru Wlut
III	Wukir	XVIII . . .	Marakih
IV	Kurantil	XIX . . .	Tambir
V	Tolū	XX . . .	Madaṅkuṅan
VI	Gumbrĕg	XXI . . .	Maha Tāl
VII	Wariga ning Wariga	XXII . . .	Wuyai
VIII	Wariga	XXIII . . .	Manahil
IX	Julung	XXIV . . .	Prang Bakat
X	Julung Sungsang	XXV . . .	Bala(muki)
XI	Duṅulan	XXVI . . .	Wugu-wugu
XII	Kuniṅan	XXVII . . .	Wayang-wayang
XIII	Laṅkir	XXVIII . . .	Kulawu
XIV	Maḍasiha	XXIX . . .	Dukut
XV	Julung Pujut	XXX . . .	Watu Gunung.

	Āditya (Sunday)	Soma (Monday)	Aṅgāra (Tuesday)	Budha (Wednesday)	Bṛhaspati (Thursday)	Çukra (Friday)	Çanaiçcara (Saturday)		
Tunglai	1	121	31	151	61	181	91	I	Pahing
Haryang	92	2	122	32	152	62	182	II	Pon
Wurukung	183	93	3	123	33	153	63	III	Wagai
Paniron	64	184	94	4	124	34	154	IV	Kaliwon
Was	155	65	185	95	5	125	35	V	Umanis
Mawulu	36	156	66	186	96	6	126	VI	Pahing
Tunglai	127	37	157	67	187	97	7	VII	Pon
Haryang	8	128	38	158	68	188	98	VIII	Wagai
Wurukung	99	9	129	39	159	69	189	IX	Kaliwon
Paniron	190	100	10	130	40	160	70	X	Umanis
Was	71	191	101	11	131	41	161	XI	Pahing
Mawulu	162	72	192	102	12	132	42	XII	Pon
Tunglai	43	163	73	193	103	13	133	XIII	Wagai
Haryang	134	44	164	74	194	104	14	XIV	Kaliwon
Wurukung	15	135	45	165	75	195	105	XV	Umanis
Paniron	106	16	136	46	166	76	196	XVI	Pahing
Was	197	107	17	137	47	167	77	XVII	Pon
Mawulu	78	198	108	18	138	48	168	XVIII	Wagai
Tunglai	169	79	199	109	19	139	49	XIX	Kaliwon
Haryang	50	170	80	200	110	20	140	XX	Umanis
Wurukung	141	51	171	81	201	111	21	XXI	Pahing
Paniron	22	142	52	172	82	202	112	XXII	Pon
Was	113	23	143	53	173	83	203	XXIII	Wagai
Mawulu	204	114	24	144	54	174	84	XXIX	Kaliwon
Tunglai	85	205	115	25	145	55	175	XXV	Umanis
Haryang	176	86	206	116	26	146	56	XXVI	Pahing
Wurukung	57	177	87	207	117	27	147	XXVII	Pon
Paniron	148	58	178	88	208	118	28	XXVIII	Wagai
Was	29	149	59	179	89	209	119	XXIX	Kaliwon
Mawulu	120	30	150	60	180	90	210	XXX	Umanis

BIBLIOGRAPHY

Basham, A. L., *The Wonder that was India*, reprint London, 1961, Appendix III: The Calendar, pp. 492-495.
Basham, A. L. (ed.), *Papers on the Date of Kaniṣka*, Leiden, 1968.
Billard, R., L'Astronomie Indienne', *Publ. E.F.E.O.*, LXXXIII, Paris, 1971.
Bosch, F. D. K., 'De inscriptie van Këloerak', *Tijdschr. Bat. Gen.*, LXVIII, 1928, pp. 1-64.
Brandes, J. L. A., 'Een rechterlijke Uitspraak uit het Jaar 927' (A Judgment of Law dated A.D. 927), *Tijdschr. Bat. Gen.*, XXXII, 1888, pp. 98-149.
——, 'Inscriptie op een zilveren Zonnescherm' (Inscription on a silver parasol), *Not. Bat. Gen.*, XXVI, 1888, pp. 20-28; reprinted in *Rapp. Oudh. Comm.*, 1911, pp. 251-259.
——, 'De maandnaam Hapit' ('Hapit' as the name of months), *Tijdschr. Bat. Gen.*, XLI, 1899, pp. 323-343.
——, 'Nog eenige Javaansche piagĕm's uit het Mohammedaansche tijdvak, afkomstig van Mataram, Bantĕn en Palembang. III. Palembang', *Tijdschr. Bat. Gen.*, XXXIV, 1891, pp. 605-623; XXXVII, 1894, pp. 121-126; XLII, 1900, pp. 131-134 and 491-507.
——, *Pararaton (Ken Arok), of het Boek der Koningen van Tumapĕl en van Majapahit* (Pararaton or Ken Arok: the Book of Kings of Tumapĕl and of Majapahit), 2nd ed. by the care of N. J. Krom with contributions by J. C. G. Jonker, H. Kraemer and R. Ng. Poerbatjaraka, The Hague-Jakarta, 1920.
——, *Oud-Javaansche Oorkonden*, edited by N. J. Krom, The Hague-Jakarta, 1913 (*Verhand. Bat. Gen.* LX).
Feuilletau de Bruyn, W. K. H., 'De Biaksche tijdrekening naar de sterrenbeelden' (Chronology in the island of Biak based on constellations), *Tijdschr. Nat. Gen.*, V, 1940-41, pp. 1-10.
Cœdès, G, 'Les inscriptions malaises de Çrīvijaya', *B.E.F.E.O.*, XXX, Fasc. 1-2, 1930, pp. 29-52.
——, 'L'origine du cycle des douze animaux au Cambodge', *T'oung Pao*, XXXI, 1934, pp. 315-329.
——, 'Les stèles de Sdok Kak Thom, Phnom Sandak et Prah Vihar (en collaboration avec P. Dupont)'. *B.E.F.E.O.*, XLIII, 1943-46, pp. 56-154.
Cunningham, A., *Book of Indian Eras with Tables for Calculating Indian Dates*, Calcutta, 1883.
Damais, L.-C., 'Études d'Épigraphie Indonésienne. I. Méthode de réduction des dates javanaises en dates européennes', *B.E.F.E.O.*, XLX, Fasc. 1, 1951, pp. 1-41; 'II. La date des inscriptions en ère de Sañjaya', *ibid.*, pp. 42-63; 'III. Liste des principales inscriptions datées de l'Indonésie', *B.E.F.E.O.*, XLVI, 1952, Fasc. 1, pp. 1-105; 'IV. Discussion des dates des inscriptions', *B.E.F.E.O.*, XLVII, Fasc. 1, pp. 7-290; 'V. Dates des manuscrits et de documents divers de Java, Bali et Lombok', *B.E.F.E.O.*, XLIX, Fasc. 1, 1958, pp. 1-239.
——, 'Études Balinaises. I. La colonnette de Sanur', *B.E.F.E.O.*, XLIV, 1951, Fasc. 1, pp. 121-128.
——, 'Études Javanaises. I. Les tombes musulmanes datées de Trâlåyå', *B.E.F.E.O.*, XLVIII, Fasc. 2, 1957, pp. 353-415.
——, 'Études sino-indonésiennes. II. Une mention de l'ère Śaka dans le "Ming Che" ', *B.E.F.E.O.*, L, 1960, pp. 30-35.
——, 'Études soumatranaises. I. La date de l'inscription de Hujuṅ Laṅit ("Bawang")', *B.E.F.E.O.*, L, Fasc. 2, 1962, pp. 275-288.
——, 'Le calendrier javanais', *Journ. Asiat.*, CCLV, Fasc. 1, 1967, pp. 133-141.
Erp, Th. van, 'Een merkwaardig hindoejavaansch chronogram in beeld in het Rijksmuseum voor Volkenkunde te Leiden' (A curious Indo-Javanese pictorial chronogram in the 'Rijksmuseum voor Volkenkunde' at Leiden), *Cult. Indië*, 1, 1939, pp. 40-47.
Fleet, J. F., 'Inscriptions of the Early Guptas', *Corpus Inscr. Indic.*, III, Calcutta, 1888, reprint Varanasi, 1970.
Fontein, J., Soekmono, R. and Suleiman, S., *Ancient Indonesian Art*, New York, 1971.
Goris, R., *Prasasti Bali*, Vols. I and II, Bandung, 1954.
Griswold, A. B., and Prasert ṇa Nagara, 'The Epigraphy of Mahādharmarāja I of Sukhodaya', *Journ. Siam Soc.*, 61, 1968, pp. 71-101.
Grohmann, A., *Arabische Chronologie*, Leiden, 1966.
Gunning, J. G. H., *Bhārata-yuddha, Oudjavaansch Heldendicht*, The Hague, 1903.
Jasper, J. E., 'Tengger en de Tenggereezen' (The Tengger and the Tenggerese), Djåwå, VI, 1926, pp. 185-192; VII, 1927, pp. 23-37, 217-231, especially, pp. 226-321.

Kohlbrugge, 'De heilige bekers der Tenggereezen' (The sacred beakers of the Tenggerese), *Tijdschr. Bat. Gen.*, XXXIX, 1897, pp. 129-141.

Krom, N. J., 'Over de dateering van eenige Kawi-geschriften' (On the dates of some Kawi texts), *Tijdschr. Bat. Gen.*, LVI, 1916, pp. 508-521.

——, *Inleiding tot de Hindoe-Javaansche Kunst*, 3 Vols., 2nd ed., The Hague, 1923.

——, *Hindoe-Javaansche Geschiedenis*, (Indo-Javanese History), 2nd ed., The Hague, 1923.

van Lohuizen-de Leeuw, J. E., *The Scythian Period*, Leiden, 1949 (especially Chapter I: The Eras (pp. 1-72).

Maass, A., 'Astrologische Kalender der Balinesen', *Feestb. Bat. Gen.*, II, 1929, pp. 126-157.

Moquette, J.-P., De eerste Vorsten van Samoedra-Pasé (Noord Sumatra)' (The first kings of Samudra-Pasai), *Rapp. Oudh. Dienst* 1913, pp. 1-12.

——, 'De oudste Mohammedaansche inscriptie op Java' (The oldest Muslim inscription in Java), *Verh. Eerste Congr. v. d. T. L. & V. van Java*, 1921, pp. 391-399.

Neugebauer, P. V., *Tafeln zur astronomischen Chronologie*, 3 pts., 2nd ed., Leipzig, 1912-1925.

Nieuwenkamp, W. O. J., 'Een Balineesche Kalender', *Bijdr. Kon. Inst.*, 69, 1914, pp. 112-126.

Noorduyn, J., 'Some aspects of Macassar-Buginese Historiography', *Historians of South East Asia*, ed. by D. G. E. Hall, London, 1961.

——, 'Origins of South Celebes Historical Writing', *An Introduction to Indonesian Historiography*, ed. by Soedjatmoko and others, Ithaca, 1965.

——, 'Tentang Asal-mulanja Penulisan Sedjarah di Sulawesi Selatan', *Madj. Ilmu-ilmu Sastra Indonesia*, III, Nomor 2 & 3, 1966, pp. 212-232.

Olthof, W. L., *Babad Tanah Djawi*, 2 Vols, The Hague, 1941.

Panchamukhi, R. S., 'The Badami Rock Inscription of Vallabheśvara, Śaka 465', *Ep. Ind.*, XXVII, 1919, pp. 4-9.

Pathak, V. S., *Ancient Historians of India. A Study in Historical Biographies*, Bombay etc., 1966.

Pigeaud, Th., 'Een stuk over sterrenkunde uit het Anggastyaparwwa', *Tijdschr. Bat. Gen.*, LXV, 1925, pp. 282-296.

——, 'Javaansche Wichelarij en Klassificatie', *Feestb. Bat. Gen.*, II, 1929, pp. 273-290.

——, *Javaans-Nederlands Handwoordenboek*, 2nd. ed., Groningen, 1945, 51 (pp. VIII-XII: 'Tijdrekening').

——, *Literature of Java*, I, The Hague, 1967, pp. 31-33 (§ 00120: Chronology).

Pocock, D. F., 'The Anthropology of Time-reckoning', *Contr. Ind. Soc.*, VII, 1964, pp. 18-29.

Porée-Maspéro, E., 'Le cycle des douze animaux dans la vie des Cambodgiens', *B.E.F.E.O.*, L, Fasc. 2, 1962, pp. 311-365.

Ricklefs, M. C., *Jogjakarta under Sultan Mangkubumi*, 1749-1792, London, 1974.

van Romondt, V. R., 'Peninggalan-peninggalan Purbakala di Gunung Penanggungan', *Laporan Dinas Purb. Rep. Indonesia*, Jakarta, 1951.

Rouffaer, G. P., 'Tijdrekening' (Chronology), *Encycl. Ned.-Indië*, IV, 1905, pp. 445-460.

Schadee, M. C., 'De Tijdrekening bij de Landak-Dajaks in de Westerafdeeling van Borneo' (Chronology among the Landak-Dayak in the Residence of West Borneo), *Tijdschr. Bat. Gen.*, LXIX, 1914, pp. 130-139.

Sewell, R. and Dikshit, S. B., *The Indian Calendar*, London, 1896.

Sircar, D. C., *Indian Epigraphy*, Delhi etc., 1965 (especially pp. 219-326).

——, *Select Inscriptions bearing on Indian History and Civilization*, I, 2nd ed., Calcutta, 1965.

Smith, R. Morton, 'On the Ancient Chronology of India', *Journ. Am. Or. Soc.*, 77, 1957, pp. 116-129 and 226-280; 78, 1958, pp. 174-192.

Stein, M. A., *Kalhaṇa's Rājataraṅgiṇī*, 2 Vols., 1900, reprint Delhi etc., 1961, especially Vol. I, pp. 56-70 (Chronology of the *Rājataraṅgiṇī*).

van Stein Callenfels, P. V., 'De historische Aji Såkå' (The historical Aji Såkå), *Tijdschr. Bat. Gen.*, LIX, 1919-21, pp. 471-479.

——, *Epigraphia Balica*, I, 1926 (*Verh. Kon. Bat. Gen.*, LXVI, 3).

Stibbe, D. G. and Spat, C., with contributions by E. M. Uhlenbeck, 'Tijdrekening' (Chronology), *Encycl. Ned.-Indië*, V, Supplement, The Hague-Leiden, 1927, pp. 401-415.

Stutterheim, W. F., *Oudheden van Bali*, 2 Vols., Bali, 1929-30, especially Vol. I (text), pp. 80-85.

——, 'Oudheidkundige Aanteekeningen. XLII: Is 1049 het sterfjaar van Erlangga?' (Archaeological Notes. XLII: Is 1049 the year of Erlangga's death?), *Bijdr. Kon. Inst.*, 92, 1935, pp. 196-202.

Teeuw, A. and others, *Śiwarātrikalpa*, The Hague, 1969 (*Bibl. Indon.* 3), especially pp. 55-67.

Teeuw, A. and Wyatt, D. K., *Hikayat Patani. The Story of Patani*, 2 Vols., The Hague, 1970 (*Bibl. Indon.* 5).

Bui Quang Tung, 'Tables synoptiques de chronologie viêtnamienne', *B.E.F.E.O.*, LI, Fasc. 1, 1963, pp. 1-78.
Vogel, J. Ph., 'The Earliest Sanskrit Inscriptions of Java', *Publ. Oudh.* Dienst, I, 1925, pp. 15-35.
van Wijk, W. E., 'On Hindu Chronology', I, *Acta Orient.*, I, 1923, pp. 206-223; II. *Ibid.*, II, 1924, pp. 55-62; III. *Ibid.*, II, 1924, pp. 235-249; IV. *Ibid.*, IV, 1926, pp. 55-80; V. *Ibid.*, V, 1927, pp. 1-27.
——, *Decimal Tables for the Reduction of Hindu Dates from the Data of the Sūrya-siddhānta*, The Hague, 1938.
Woelders, M. O., *Het Sultanaat Palembang*, 1811-1825, The Hague, 1975 (*Verh. Kon. Inst.*, 72).
Zoetmulder, P. J., *Kalangwan. A Survey of Old Javanese Literature*, The Hague, 1974 (*Kon. Inst., Transl. Series* 16).

INDEX

(All figures refer to pages; additional asterisks indicate footnotes)

ā, abbreviation of Āditya(wāra), 3
Abhijit, name of the 28th *nakṣatra* in the system of Brahmagupta, 24, 24*
Aceh (Atjeh, Acheen), 35, 42
Āditya-wāra, 1st day of the seven-day week: Sunday, 3
āgneya(-maṇḍala), the southeast(ern section of the sky), 23*
Ahad (Akad), Sunday, 4
Airlaṅga (Erlangga), Old Javanese king (*c.* 1019-1042), 27, 27*, 28*, 30
alip, 1st year of the Javanese eight-year cycle, 39*
amānta, of months: ending in, and therefore also beginning with, a new moon, 7
Amoghapāśa, inscribed stone statue from Padang Roco (Sungai Langsat, Central Sumatra), 24 f.
ang, abbreviation of Aṅgāra(wāra), 3
Aṅgāra-wāra, 3rd day of the seven-day week: Tuesday, 3
Angkor Borei (Cambodia), inscription of, 16 f.
animals cycle of twelve, 35, 35*
anumoda, 'pious donation', 32
Ārdrā, name of the 6th *nakṣatra*, 17*
Argapura (Central Java), inscription of, 22*
Arjunawiwāha, name of one of the oldest datable Old Javanese texts, 27, 27*, 40
Arjuno, see s.v. Gunung Arjuno
Āryabhaṭa, ancient Indian astronomer, 7*
Āryabhaṭīya or *Āryasiddhānta*, ancient Indian astronomical text composed by Āryabhaṭa, 7, 7*
Āṣāḍha, 4th month of the ancient Indian and Indonesian year, 39, 40*, 48.
Aśoka, inscriptions of, 12
astronomy, significance of detailed astronomical data in inscriptions, 18 f.
Asuji(-māsa), usual Old Javanese form of the Indian month Āśvina or Āśvayuja (the 7th of the year), 27, 48.
Azes I, Indo-Parthian king regarded by some as the founder of the Vikrama Era, 9

Babad ing Sĕngkala, dates in the, 44
Babad Tanah Jawi, dates in the, 44, 44*, 46
Badami, rock inscription of Vallabheśvara, 14, 14*
Bala, name of the 25th *wuku*, 4*
Bali, dating of inscriptions from, 25, 25*, 26; 'zodiac beakers' from, 32
Balinese, 3, 3*, 18 etc.
Balitung, dating of inscriptions of, 17*, 23*
Bāṇa, ancient Indian writer and historian, author of the *Harṣacarita* etc., 27, 27*
Bandhuvarman, Mandasor inscription of, 28*
Bantĕn, 42, 44*
Baruṇa, *bāruṇya* (*-maṇḍala*), the west(ern section of the sky), 23*
Barus, Tamil inscription of, 11
Basak, R. G., 16*
Basham, A. L., 6*, 9*, 47
Batutulis (or Batu Tulis) near Bogor, West Java, stone inscription of, 26, 26*
bava (*bawa*), name of a *karaṇa*, 23
bāyabya (correct Sanskrit spelling: *vāyavya*), the north-east(ern section, *maṇḍala*, of the sky), 23*
be, (1) abbreviation of Beteng, 3; (2) 6th year of the Javanese eight-year cycle (*windu*), 39*, 44
Bĕlahan, pictographic chronogram at, 30, 30*
Bengkahulu, see s.v. Hujung Langit
Berg, C. C., 40*
Beteng, 2nd day of the Balinese three-day week, 3
Bhāratayuddha, Old Javanese *kĕkawin* dated A.D. 1157, 27, 27*
bhasmi-bhuta, Sanskrit *bhasmībhūta*, 'reduced to ashes', but the 1st part of the compound also has the numerical value of 0, the 2nd that of 5 (*pañcamahābhūta*, the five elements), 29
bhumi, Sanskrit *bhūmi*, 'earth', also used to express the figure 1.
Bhaṭāra Guru, 32*
Billard, R., 6*, 10*, 47,
Biluluk, charters of, 20*
Bosch, F. D. K., 15*, 28*, 32*, 47
Brahmagupta, name of ancient Indian astronomer, 7*, 24, 24*
Brahmasiddhānta or *Brahmasphuṭasiddhānta*, ancient Indian text on astronomy, 7, 7*, 23
Brandes, J. L. A., 2, 2*, 19*, 22*, 28*, 29*, 39*, 40, 40*, 43*, 44*, 47
brĕ (*wrĕ*), abbreviation of Brĕhaspati(wāra)
Brĕhaspati-wāra, Thursday, 3, 4*, 21
Bṛhajjātaka, ancient Indian text on astronomy, 22
Brown, C. C., 43*
bu (*wu*), abbreviation of Budha(-wāra)
Budha-wāra, 'Wednesday', 3, 21
Bugis, Buginese, diaries, 45
Bühler, G., 11*
bulan, 'moon, month' in Indonesian etc., 5, 25
Byantara, 3rd day of the three-day week in Bali, 3*

Haryang (Hariyang), 2nd day of the six-day week, 3
Hathigumpha (Puri Distr., Orissa, India), cave inscription of Khāravela, 12
Hijrah era, 24, 33-38, 41 f.
Hikayat Banjar, absence of dates in the 43, 43*
Hikayat Malèm Dagang, absence of dates in the, 43, 43*
Hooykaas, C., 35*
Hujung Langit (Krui, Bengkahulu, South Sumatra), inscription of, 24
Hultzsch, E., 11*

Indian civilization, expansion of, 15, 17
indriya, 'senses', used to express the figure 5, 29

Jains, Jainism, 14, 14*, 15
Jalatunda, rock-cut date at, 31, 31*
Jasper, J. E., 47
Java, Javanese, *passim*
Jayabhaya, king of Kaḍiri, East Java (*c.* 1135-1157 A.D.), 22*
Jayanagara, king of Majapahit (1309-1328), 29
je, 4th year of the Javanese eight-year cycle (*windu*), 39*
jimakir, 8th year of the Javanese eight-year cycle (*windu*), 39*
jimawal, 3rd year of the Javanese eight-year cycle (*windu*), 39*
Jonker, J. C. G., 29*, 43*
Jovian (*bārhaspatya*), see s.v. year
Jum'at, 'Friday' of the Muslim week
Junagarh (Kathiawar, India), rock inscription of Skanda Gupta, 28*
Jupiter, cf. s.v. *grahacāra* and Bṛhaspati
Jyaiṣṭha or Jyeṣṭha, 3rd month of the ancient Indian and Indonesian year, 8, 16, 39, 40*, 48.

ka, abbreviation of Kaliwuan, 3; abbreviation of Kajeng, 3
Kajeng, 3rd day of the three-day week in Bali, 3
Kāla, 'Time', also identified with Śiva as the Divine Destroyer, 6
Kalacuri era, 9, 12, 12*
Kālakācāryakathā(naka), 14*
Kalasan, Sanskrit inscription of, 28*
Kaliwuan, 4th day of the five-day week, 3, 27 (also Kaliwon)
Kaliyuga (era), 11, 11*, 12, 30
kalpa, the longest cycle of time in Indian cosmology, also defined as a 'day of Brahmā', equated with 4,320 million years (A. L. Basham, *The Wonder that was India*, London, 1954, p. 320)
Kane, P. V., 4*, 6*, 18*, 22*, 33*
Kāṇwa, Pu, author of the *Arjunawiwāha*, 27, 40
kanyādi, in ancient India: solar years starting at the time when the sun enters the sign of Virgo, 41
karaṇa, half of a *tithi*, 18, 23, 23*, 25
Karkaṭa(-rāśi), the zodiac sign Cancer, 22, 54
Kārttika (often spelt Kārtika, although the *-tt-* is here organic, as the word is derived from *Kṛttikā*, 'the Pleiades'), 8th month of the ancient Indian and Indonesian year, 27, 48
kaulava, name of the 4th *karaṇa*, 23
Kĕdukan Bukit (Palembang, South Sumatra), inscription of, 17
kĕkawin, Old Javanese poetical composition in Indian metres, 27, 27*, 28*
Kĕlurak, Sanskrit inscription of, 28*
Kerala, 41, 41*
Kern, H., 12*, 29*
Khāravela, see s.v. Hāthigumphā
Khmer, Old Khmer inscriptions, 17, 17*
kidung, Old Javanese poetical composition in Indonesian metres, 27, 27*
Kielhorn, F., 12*
kiṃstughna, name of the 1st *karaṇa* of the month, 23
Kliyon, see s.v. Kaliwuan
Kohlbrugge, J. H. F., 32*, 48
koḷḷam āṇḍu, see s.v. eras
Kra, Isthmus of, 35
Kraemer, H., 29*, 43*
Krishna Sastri, 16*
Krom, N. J., 19, 29, 29*, 31*, 32*, 43*, 48
kṛṣṇa-pakṣa, 2nd half of the month, during the waning moon from the full to the new moon, 9
Kṛta era, 12*
Kṛtarājasa, Pĕnanggungan copper-plate inscription of, 19, 19*
Krui, see s.v. Hujung Langit
Kumāra Gupta, 28*
Kumbha(-rāśi), name of the zodiac sign Aquarius, 22*, 54
Kuniṅan, name of a *wuku*, 24
Kuṣāṇas, name of a dynasty of ancient India; mention of *nakṣatra* in inscriptions of, 17

lagna, 17*, 22, 22*
Lamb, A., 11*
Landak-Dayaks (West Kalimantan), chronology of, 43
Laukika era, also called Saptarṣi era, 11, 11*, 12
Leran (Lèran, Rembang, Central Java), tomb stone of, 34, 34*
Lobok Tua (near Barus, Tapanuli, North Sumatra), Tamil inscription of, 11, 11*
Lohuizen-de Leeuw, J. E. van, 10*
Lokavibhāga, dated Jain text, 14, 14*
Lulius van Goor, M. E., 31*

ma, abbreviation of Mawulu, 3
Maass, Alfred, 18*, 48
Macassar, diaries, 45
Maḍangkungan, name of a *wuku*, 27

Mādhava, name of a Vedic month, corresponding to Vaisākha in later times, 40
Madhu, name of a Vedic month, corresponding to Caitra in later times, 40, 40*
Madura, island of, 30
Māgha(māsa), 10th month of the ancient Indian and Indonesian year, 17*, 48
mahā-, before the names of months to indicate years of the 12-year 'Jovian' cycle, 35
mahendra(-maṇḍala), the east(ern section of the sky), 23*
makara(-saṃkrānti), entrance of the sun into the zodiac sign of Cancer, sometimes deified as a form of Durgā, 33*, 54
Malacca, 42
Maliku's-Saleḥ, name of the first Sultan of Samudra-Pasai; his tomb stone, 34, 34*
Malleret, L., 17*
maṇḍala, each of the eight sectors of the sky, 21, 22, 23, 23*
Mandasor, stone inscription of Kumāra Gupta and Bandhuvarman, 28*
mangsa, 39
maṅilala drawya haji, 19*
Marrison, G. E., 35*
māsa, see s.v. month
Mataram, dated *piagĕms* from, 44*
Mawulu, 6th day of the six-day week, 3, 27
Maymūn, daughter of 34
Millies, 32*
mīna(-saṃkrānti), entrance of the sun into the zodiac sign Pisces, 8; True Mīna Saṃkrānti (T.M.S.), with first new moon after this marking the beginning of the new year in one of the most common ancient Indian and Indonesian systems, 40*, 54
Minggu, normally preceded by *hari* in Indonesian, Sunday, 4*
Minto Stone, inscription of king Wawa (A.D. 928), 19
Minye Tujuh (Aceh), inscription of, 35, 35*
Mirashi, V. V., 12*, 16*
mithuna, the zodiac sign Gemini, represented in Indonesia by paired crabs, 33*, 54
month, synodic, 7; intercalation of, 7 f.; sidereal, 9; solar (*saura*), 13, 13*; ending at full moon (*pūrṇimānta*), 13, 13*, 14; ending at new moon (*amānta*), 14; Muslim, 34, 37; Javanese solar, 39, 40; 48
Moquette, J.-P., 34*, 48
Myinpagan near Pagan, Burma, Tamil inscription of, 11, 11*

nāga, (1) name of the last but one *karaṇa* of the month, 23; (2) used to express the figure 8, 32, 32*
Nāgarakrĕtāgama, Old Javanese text dated A.D. 1365; dates in the, 30, 43
nakṣatra, 'lunar mansion', each of the 27 or 28 divisions of a sidereal month of 27, 322 days and corresponding to 1,012 days according to the most common (equal-time) system, 5*, 9, 17, 18, 21, 22, 22*, 23, 24, 24*, 52
navagraha, the nine 'planets', *viz.* the five (Mercury, Venus, Mars, Jupiter and Saturn), as well as the sun and the moon, and the ascending and descending nodes (Rāhu and Ketu), 30; cf. s.v. Rāhu and *grahacāra*
Neugebauer, P. W., 48
Ngabean, copper-plate inscription of Kayuwangi, 19
Nieuwenkamp, W. O. J., 3*, 42, 42*, 48
Nilakanta Sastri, K. A., 11*, 13*, 14*
Nirvāṇa era, 9
Noorduyn, J., 45, 45*, 48

Old Balinese, 3
Old Javanese, *passim*
Old Malay, 17
Olthof, W. L., 44*, 48

pa, (1) abbreviation of Pahing, 3; (2) of Paniruan, 3; (3) of Pasah, 3
Padang Roco, see s.v. Amoghapāśa
Pagan, Tamil inscription of, 11, 11*
Pahang, name of a *wuku*, 20*, 27
Pahing, 1st day of the five-day week, 3
Pajajaran, old Sundanese kingdom; foundation stone of, 26*
pakṣa, each of the two halves of a synodic lunar month: the waxing (*śuklapakṣa*) and the waning moon (*kṛṣṇapakṣa*, which are each divided into 15 *tithis*, 9, 17, 25
Pālas, ancient royal dynasty of Bengal and Bihar (from the 8th to the 12th century), 15
palawija, crops other than rice, 40*
Palembang (South Sumatra), *piagĕms* from, 44, 44*, 45, 45*; see also s.v. Kĕdukan Bukit
Pallavas, ancient South Indian royal dynasty (from the 4th to the 9th century A.D.), 14, 16*
panakawan, 33*
Panataran, dates on various buildings of the complex, 31 f.
pañcāṅga, 18, 23
pañca-wāra, five-day week; cf. s.v. week
Panchamukhi, R. S., 14*, 48
Pāṇḍyas, ancient Indian royal dynasty; dating of their inscriptions, 14*
Paniruan (Panirwan, Paniron), 4th day of the six-day week, 3
Panuluh, see s.v. *Bhāratayuddha*
Pararaton, dates in the, 29
Pasah, 1st day of the 3-day week in Bali, 3
Pasai, cf. s.v. Maliku-'s-Salih
Pathak, V. S., 48
Pĕnanggungan, mountain in East Java; inscription of the, 19; pictorial chronogram at Bĕlahan, 30; dates on different monuments, 31

LIST OF ILLUSTRATIONS

All illustrations are published by courtesy of the Kern Institute, Leiden, Netherlands, except for Plate I, No. 2, which is based on a photograph in the collection of Mrs. H. I. R. Hinzler-Schotermans, who kindly gave me permission to publish her photograph of a fragment which was hitherto unknown or, at least, unrecorded.

PLATE II

1

2

3

Plate IV

Printed in the United States
By Bookmasters